NELSON MATHS

AUSTRALIAN CURRICULUM

Student Book

Glenda Bradley

Contents

Writing Numbers to 5

DATE:

✿ Trace over the numbers.

1 🐝	1	1	1
2 🐝🐝	2	2	2
3 🐝🐝🐝	3	3	3
4 🐝🐝🐝🐝	4	4	4
5 🐝🐝🐝🐝🐝	5	5	5

✿ Trace over the numbers.

Unit 1

Numbers to 5 (TRB pp. 22–25)
Number and place value Establish understanding of the language and processes of counting by naming numbers in sequences, initially to and from 20, moving from any starting point (ACMNA001) **AC**

Bingo 5

✿ Play a game.

You will need:

- a dice with a ★ sticker covering the 6
- a partner 😊

Make your bingo grid:

- Roll the dice and say the number to your partner.
- Your partner writes the number in a box below.
- Take turns rolling the dice and saying the number until every box has a number in it.
- If you roll a ★, you can choose 1 2 3 4 or 5.

How to play:

- In turn, roll the dice.
- Say the number. If you have that number in a box, colour it in.
- When you have coloured 5 numbers in a row, you win. Good luck!

Unit 1 **Numbers to 5** (TRB pp. 22–25)
Number and place value Establish understanding of the language and processes of counting by naming numbers in sequences, initially to and from 20, moving from any starting point
(ACMNA001) AC

5

Dot-to-Dot

Join the numbers in the correct order.

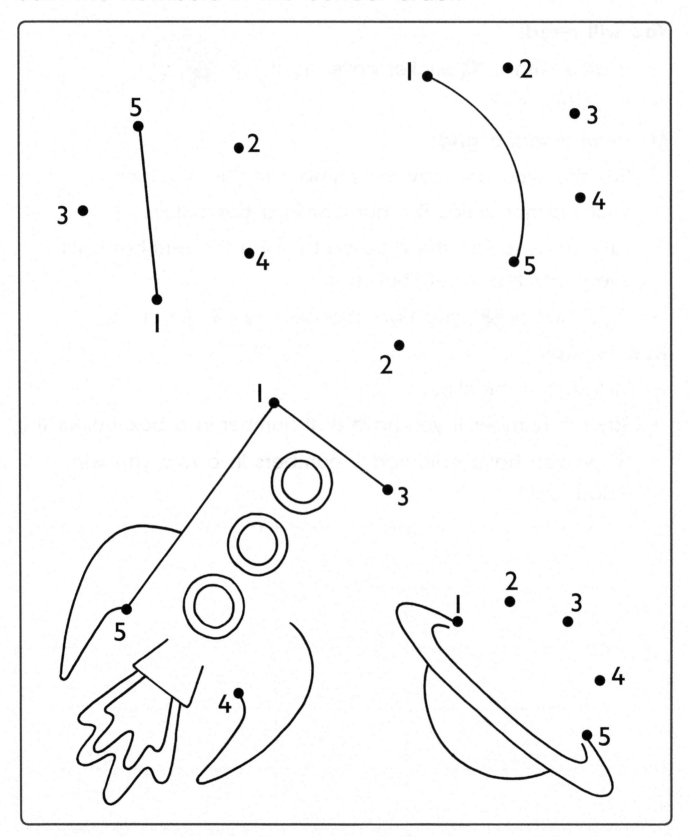

Unit 1
Numbers to 5 (TRB pp. 22–25)
Number and place value Establish understanding of the language and processes of counting by naming numbers in sequences, initially to and from 20, moving from any starting point (ACMNA001) AC

STUDENT ASSESSMENT

✿ Count up to 5.

✿ Write the numbers in the correct order.

3   5   2   4   1

Write these numbers in the correct order, too.

5   3   1   4   2

✿ Start at 1 and join the dots.

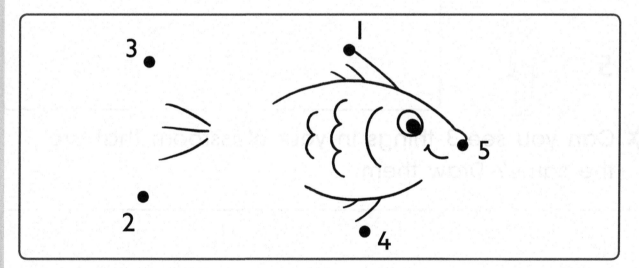

✿ Ask your partner to tell you a number.

Write it in the box.

✿ Have your partner check your number.

Unit 1

Numbers to 5 (TRB pp. 22–25)
Number and place value Establish understanding of the language and processes of counting
by naming numbers in sequences, initially to and from 20, moving from any starting point
(ACMNA001) AC

7

In the Classroom

✿ In the big box, draw:

2

4

1

3

5

✿ Can you see 3 things in your classroom that are the same? Draw them.

Unit 2
Counting to 5 (TRB pp. 26–29)
Number and place value Connect number names, numerals and quantities, including zero, initially up to 10 and then beyond
(ACMNA002) AC

School Things

✭ Draw the correct number of things in the box.

3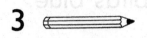

1

4

2

5

✭ Draw a circle around the number that matches the group.

1   2   3   4   5

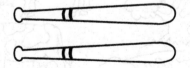

1   2   3   4   5

Unit 2

Counting to 5 (TRB pp. 26–29)
Number and place value Connect number names, numerals and quantities, including zero, initially up to 10 and then beyond
(ACMNA002) **AC**

9

At the Farm

✸ Colour 2 horses brown.

Colour 4 pigs pink.

Colour 1 sheep purple.

Colour 5 cows black.

Colour 1 fish orange.

Colour 3 hens red.

Colour 4 ducks yellow.

Colour 2 birds blue.

Unit 2 **Counting to 5** (TRB pp. 26–29)
Number and place value Connect number names, numerals and quantities, including zero, initially up to 10 and then beyond
(ACMNA002) **AC**

STUDENT ASSESSMENT

✡ Look at the bowl below.

How many  ? _____

How many 🍌 ? _____

✡ Draw 4 🍓 and 2 🍎 and 1 🍐 in the bowl.

✡ Colour 3 of the 🫐 green.

✡ Colour 2 of the 🍌 yellow.

✡ What other fruit can you put in the bowl?

Draw the fruit.

Unit 2

Counting to 5 (TRB pp. 26–29)
Number and place value Connect number names, numerals and quantities, including zero, initially up to 10 and then beyond
(ACMNA002) AC

11

Number Words

✦ Complete the titles of the books by writing the number word.

✦ Draw:

four 🦆	one 🐢
three 🐍	five 🐱
two 🪲	three 🦋

12 **Unit 3** **Groups of Things** (TRB pp. 30–33)
Number and place value Connect number names, numerals and quantities, including zero, initially up to 10 and then beyond
(ACMNA002) AC

Compare, order and make correspondences between collections, initially to 20, and explain reasoning
(ACMNA289) AC

All About Four

✦ Colour the groups that have **four** things.

How many groups did you colour? _____

✦ Can you find more groups of four in
your classroom?

Draw them.

Unit
3
Groups of Things (TRB pp. 30–33)
Number and place value Connect number names, numerals
and quantities, including zero, initially up to 10 and then beyond
(ACMNA002) AC

Compare, order and make correspondences between
collections, initially to 20, and explain reasoning
(ACMNA289) AC

13

Less and More

Draw a group that is less.		Draw a group that is more.
	(4 pencils)	
	(3 scissors)	
	(2 backpacks)	
	(3 FISH books)	
	(4 children)	

Now draw your own group in the middle.

Draw a group that is less.		Draw a group that is more.

14   Unit 3   **Groups of Things** (TRB pp. 30–33)
Number and place value Connect number names, numerals
and quantities, including zero, initially up to 10 and then beyond
(ACMNA002) AC

Compare, order and make correspondences between
collections, initially to 20, and explain reasoning
(ACMNA289) AC

DATE:

STUDENT ASSESSMENT

✸ Finish the table.

1	one	
4		
	two	
		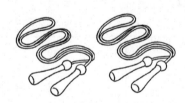
5		

✸ Colour the groups that have the **same** number of things.

✸ Look at the fish.

Draw a group that is **more**.

 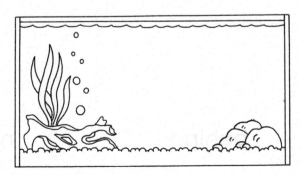

Unit 3 **Groups of Things** (TRB pp. 30–33)
Number and place value Connect number names, numerals and quantities, including zero, initially up to 10 and then beyond (ACMNA002) **AC**

Compare, order and make correspondences between collections, initially to 20, and explain reasoning (ACMNA289) **AC**

15

Where Am I?

✿ Write a word to match each picture.

near   between   next to   in front of   behind   under

_____   _____   _____

✿ Draw a picture for each word.

behind	near	in front of

16   **Unit 4**   **Location** (TRB pp. 34–37)
Location and transformation Describe position and movement
(ACMMG010) **AC**

Where Did Rosie Walk?

✪ Show where Rosie walked.

✪ Use the words to label how she walked.

past	around	under	across	over	through

Monkey Trouble

Draw things for Baby Monkey to go near, across, between, under and next to on her way back to Mother.

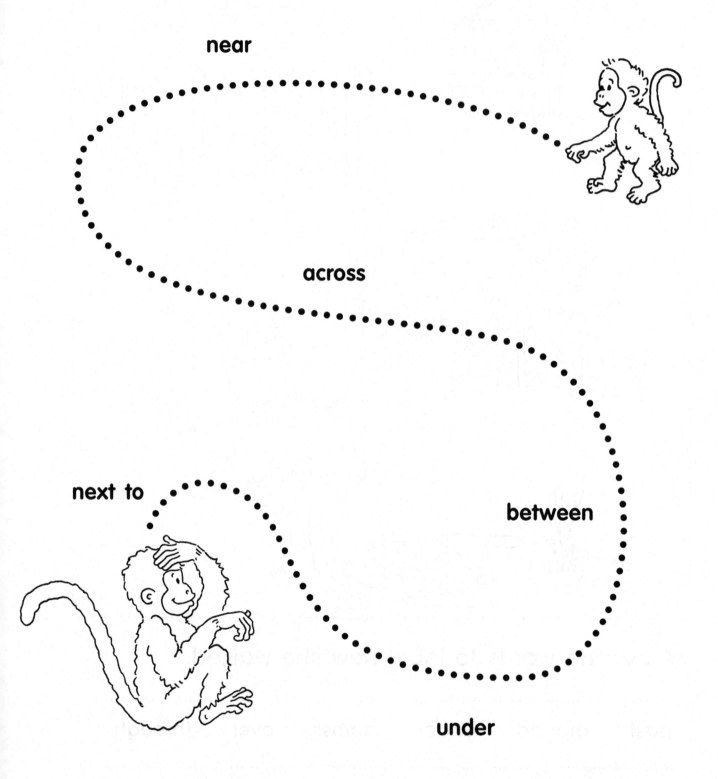

near

across

next to

between

under

STUDENT ASSESSMENT

✸ Draw more farm animals in the correct places.

A sheep is **between** the horse and the cow.

A pig is **next to** the horse.

2 ducks are **near** the cow.

✸ Draw a farmer and the path she takes.

First, she walks **towards** the pig.

Then she walks **between** the horse and the sheep.

Then the farmer walks **forwards** to the ducks.

At last she stands **next to** the cow.

Rocket Blast-Off!

✦ Finish the poem. Use the numbers 5, 4, 3, 2, 1, 0.

Zoom! Zoom! Zoom!
We're going to the moon
Zoom! Zoom! Zoom!
We're going to the moon
We'll climb aboard a rocket ship
And go upon a little trip
Zoom! Zoom! Zoom!
We're going to the moon

_____, _____, _____, _____, _____, _____, Blast-off!

✦ Play a game.

You will need:

- a dice with the 6 changed to a 0 🎲
- a partner 🙂

How to play:

- In turn, roll the dice.
- Write the number you rolled on your rocket. Make sure it is in the correct order.
- When you have filled in 5, 4, 3, 2, 1, 0, colour in the "Blast-off".
- The first player to colour their "Blast-off" wins.

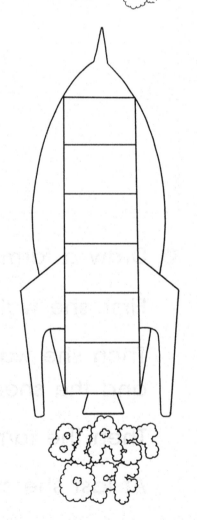

20 Unit 5 **More Counting** (TRB pp. 38–41)
Number and place value Establish understanding of the language and processes of counting
by naming numbers in sequences, initially to and from 20, moving from any starting point
(ACMNA001) AC

Before

✪ Write the number that comes **before**:

_____ 6   _____ 3   _____ 1   _____ 4   _____ 5   _____ 2

✪ Play a game.

You will need:

- some counters, a dice and a partner

How to play:

- Put your counters on START and roll the dice.
- Move to the number that comes **before** the number you rolled.
- The first player to pass FINISH wins.

Unit 5 **More Counting** (TRB pp. 38–41)
Number and place value Establish understanding of the language and processes of counting by naming numbers in sequences, initially to and from 20, moving from any starting point
(ACMNA001) (AC)

21

Missing Numbers

✿ Write the missing numbers.

0 1   3 4 5 6

0 1 2 3   5 6

6 5   3 2 1 0

✿ Write your own numbers.

✿ Write the missing numbers.

0

1

6

7 8

9

Unit 5   **More Counting** (TRB pp. 38–41)
Number and place value Establish understanding of the language and processes of counting
by naming numbers in sequences, initially to and from 20, moving from any starting point
(ACMNA001) AC

DATE:

STUDENT ASSESSMENT

✶ Complete the table.

Write the number that comes **before**.		Write the number that comes **after**.
	2	
	4	
	1	
	5	
	3	

✶ Count backwards from 6.

6 _____

✶ Fill in the missing numbers.

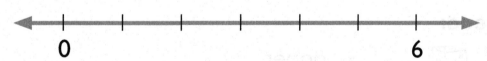

0   6

✶ Some shirts are missing. Draw them.

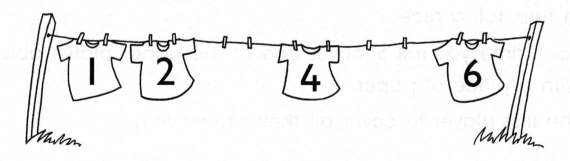

Unit
5
More Counting (TRB pp. 38–41)
Number and place value Establish understanding of the language and processes of counting
by naming numbers in sequences, initially to and from 20, moving from any starting point
(ACMNA001) **AC**

23

Dot Plates

✪ Show 5 in different ways. Use a paper plate and some counters.

Draw some of the ways you found.

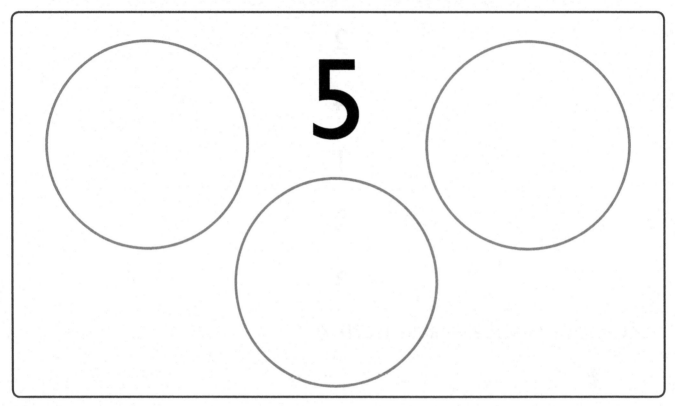

✪ Play a game.

You will need:

- a dice
- paper
- a partner

How to play:

- In turn, roll a dice.
- Each time you roll 5, cover one of the paper plates above with a piece of paper.
- The first player to cover all their plates wins.

24   Unit 6   **Dot Patterns** (TRB pp. 42–45)
Number and place value Subitise small collections of objects
(ACMNA003) AC

Glub! Glub!

✡ Draw dots on each frog to match the number of bugs it will eat. One has been done.

✡ On the lily pad, write the matching number word. One has been done.

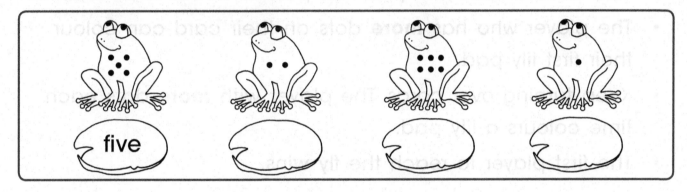

five

✡ Draw a line matching each frog to a lily pad and a bug. One has been done.

five   six   one   two

Hungry Frogs

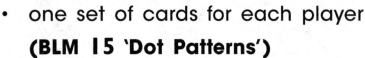

Play a game.

You will need:

- a partner
- one set of cards for each player
  (BLM 15 'Dot Patterns')

How to play:

- Choose a frog. Turn all your cards over and place them in a pile.

- Take a card from your pile. Your partner needs to do this, too.

- The player who has **more** dots on their card can colour their first lily pad.

- Keep turning over cards. The player with **more** dots each time colours a lily pad.

- The first player to reach the fly wins.

STUDENT ASSESSMENT

✪ Help Tommy Turtle get to the letterbox.
Colour in the stepping stones that show 5.

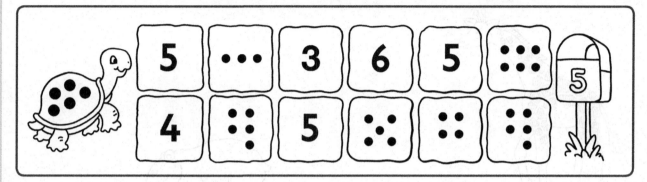

✪ Draw a friend for
Tommy Turtle who has
more dots than Tommy.

✪ Draw a friend for
Tommy Turtle who has
less dots than Tommy.

✪ Draw a friend with
the **same** number
of dots as Tommy Turtle.
Use a different pattern.

At the Zoo

✿ Draw a home for each animal. Make sure each home is **taller than** the animal.

✿ Colour the **tallest** animal yellow.

✿ Colour the **shortest** animal blue.

Length (TRB pp. 46–49)
Using units of measurement Use direct and indirect comparisons to decide which is longer, heavier or holds more, and explain reasoning in everyday language

(ACMMG006) AC

Pencils

✿ Colour the **longest** pencil green.

✿ Look around the classroom. Draw something that is **longer than** the longest pencil.

✿ Colour the **shortest** pencil yellow.

✿ Look around the classroom. Draw something that is **shorter than** the shortest pencil.

✿ Draw a pencil that is **shorter than** the longest pencil but **longer than** the shortest pencil.

Unit 7 **Length** (TRB pp. 46–49)
Using units of measurement Use direct and indirect comparisons to decide which is longer, heavier or holds more, and explain reasoning in everyday language
(ACMMG006) AC

29

Snakes Alive

✦ Colour the **longest** snake blue.

✦ Colour the **shortest** snake red.

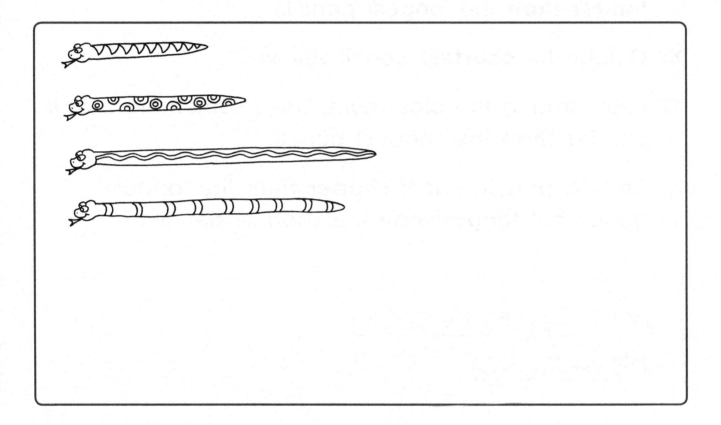

✦ Cut a piece of string that is the **same** length as the blue snake.

✦ Look around the classroom. Draw something that is **shorter than** the blue snake.

✦ Look around the classroom. Draw something that is **longer than** the blue snake.

✦ Look around the classroom. Draw something that is the **same** length as the blue snake.

30   Unit 7   **Length** (TRB pp. 46–49)
Using units of measurement Use direct and indirect comparisons to decide which is longer, heavier or holds more, and explain reasoning in everyday language
(ACMMG006) AC

STUDENT ASSESSMENT

✡ Look around the classroom. Draw something that is **longer than** your pencil.

✡ Draw something that is **shorter than** your shoe.

✡ Colour the **longest** car red.

✡ How do you know which car is longer?

_____

_____

Length (TRB pp. 46–49)
Using units of measurement Use direct and indirect comparisons to decide which is longer,
heavier or holds more, and explain reasoning in everyday language
(ACMMG006) **AC**

31

A Game of Numbers

Play a game.

You will need:

- a 10-sided dice
- a partner

How to play:

- In turn, roll the dice and trace over the number you rolled.
- The first person to trace over all their numbers wins.

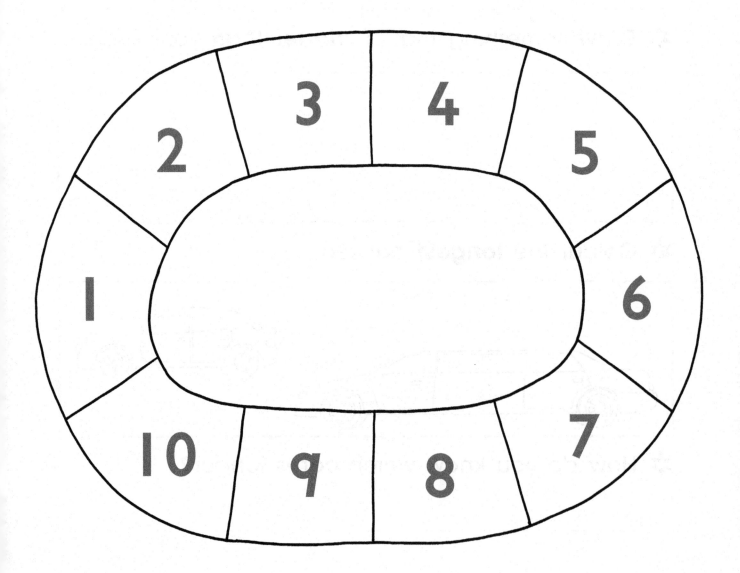

32 Unit 8 **Numbers to 10** (TRB pp. 50–53)
Number and place value Connect number names, numerals and quantities, including zero, initially up to 10 and then beyond
(ACMNA002) AC

Subitise small collections of objects
(ACMNA003) AC

Stepping Stones

✪ Colour the stepping stones to help the koala cross the river. You must colour **all** the stones from 1 to 10.

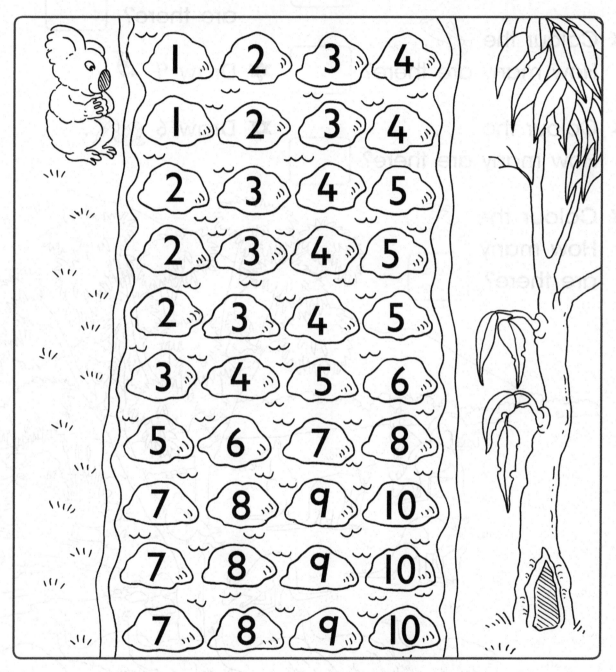

✪ Write the numbers in the correct order.

6   1   9   3   4   7   8   10   5   2

Unit **8** **Numbers to 10** (TRB pp. 50–53)
Number and place value Connect number names, numerals and
quantities, including zero, initially up to 10 and then beyond
(ACMNA002) **AC**

33

In the Bush

✸ Colour the  .
How many are there? ☐

✸ Colour the 🐟 .
How many are there? ☐

✸ Colour the 🐨 .
How many are there? ☐

✸ Colour the 🦎 .
How many are there? ☐

✸ Colour the 🦤 .
How many are there? ☐

✸ Draw 9 🐦 .

✸ Draw 6 🐍 .

34 Unit 8 **Numbers to 10** (TRB pp. 50–53)
Number and place value Connect number names, numerals
and quantities, including zero, initially up to 10 and then beyond
(ACMNA002) AC

DATE:

STUDENT ASSESSMENT

✪ Write the numbers from 1 to 10.

✪ How many?

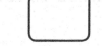

✪ Draw:

4

9

7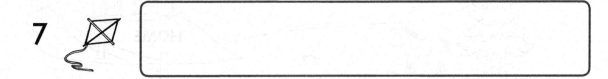

✪ Write the numbers in the correct order.

2   3   4   6   5   7   8   10   9

Unit
8
Numbers to 10 (TRB pp. 50–53)
Number and place value Connect number names, numerals
and quantities, including zero, initially up to 10 and then beyond
(ACMNA002) **AC**

Subitise small collections of objects
(ACMN003) **AC**

35

Are We There Yet?

✡ Count backwards.

10 _____ 0

✡ Write the numbers in the correct order to help the family get home.

✡ Play a game.

You will need: a 10-sided dice and a partner

• In turn, roll the dice until you roll a 10.

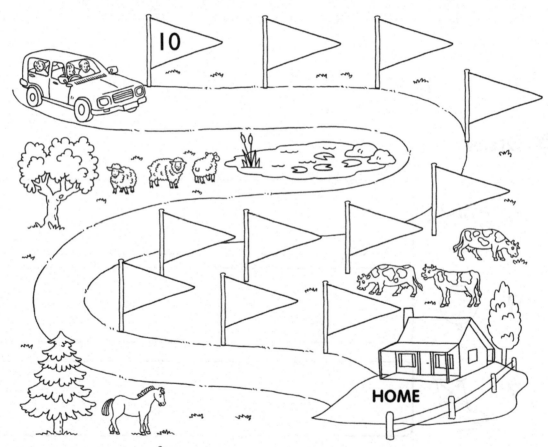

• Now colour in the ⚑.

• Keep rolling the dice until you get a 9. Colour in the ⚑.

• Keep rolling the dice until you have coloured all the signposts and the family is home ⚑.

36 Unit 9 **Counting with Numbers to 10** (TRB pp. 54–57)
Number and place value Establish understanding of the language and processes of counting by naming numbers in sequences, initially to and from 20, moving from any starting point
(ACMNA001) AC

Sorting Cards

You will need: a 10-sided dice

✿ Sort the cards starting at 1.

1									

✿ Sort again starting at 10.

10									

✿ Throw the dice. Write the number in the box.

Now count **forwards** as far as you can and write the numbers.

✿ Throw the dice. Write the number in the box.

Now count **backwards** as far as you can and write the numbers.

Unit 9 **Counting with Numbers to 10** (TRB pp. 54–57)
Number and place value Establish understanding of the language and processes of counting by naming numbers in sequences, initially to and from 20, moving from any starting point
(ACMNA001) **AC**

37

A Different Dot-to-Dot

✿ Join the dots.

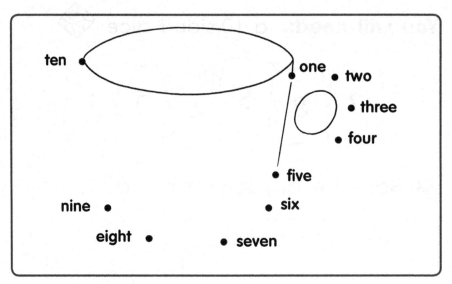

✿ Draw a line matching the pictures to the words.

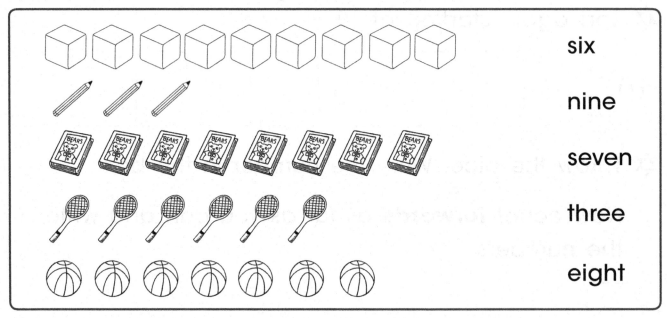

six

nine

seven

three

eight

✿ Draw a group of things for each number.

seven	four
ten	six
eight	five

Unit 9

Counting with Numbers to 10 (TRB pp. 54–57)
Number and place value Establish
understanding of the language
and processes of counting by

naming numbers in sequences,
initially to and from 20, moving
from any starting point
(ACMNA001) **AC**

Connect number names, numerals
and quantities, including zero,
initially up to 10 and then beyond
(ACMNA002) **AC**

STUDENT ASSESSMENT

� Finish the counting patterns.

10   9   8   _____

5   _____

8   7   _____

� Start counting up from the number on the dice.

1   _____ _____ _____

5   _____ _____ _____

2   _____ _____ _____

� Draw a line to match the cards.

three		6
nine		7
six		3
seven		9

� Draw a group of things for each number.

five	seven
ten	eight

Unit 9 Counting with Numbers to 10 (TRB pp. 54–57)
Number and place value Establish
understanding of the language
and processes of counting by   naming numbers in sequences,
initially to and from 20, moving
from any starting point
(ACMNA001) AC   Connect number names, numerals
and quantities, including zero,
initially up to 10 and then beyond
(ACMNA002) AC   39

Using Ten Frames

✡ Show 8 on the ten frame.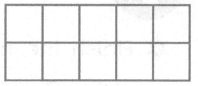

How many empty spaces? _____

✡ How many dots? _____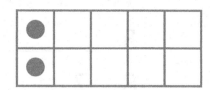

How many empty spaces? _____

✡ How many dots? _____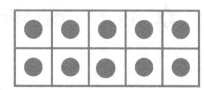

How many empty spaces? _____

✡ How many dots? _____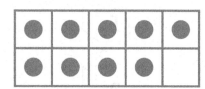

How many empty spaces? _____

✡ How many dots? _____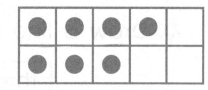

How many empty spaces? _____

✡ How many dots? _____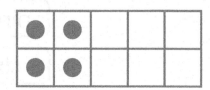

How many empty spaces? _____

✡ How many dots? _____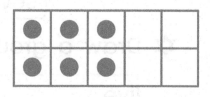

How many empty spaces? _____

✡ Show 6 another way.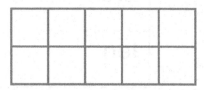

Ten Frames (TRB pp. 58–61)
Number and place value Connect number names, numerals
and quantities, including zero, initially up to 10 and then beyond
(ACMNA002) **AC**

Subitise small collections of objects
(ACMNA003) **AC**

Ten Frame Trains

Complete the trains. One has been done.

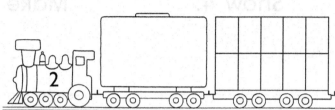

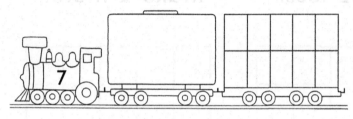

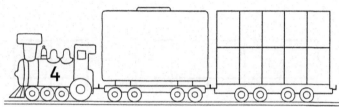

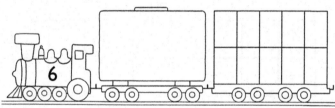

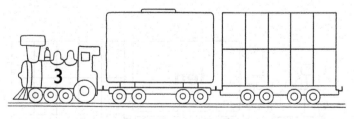

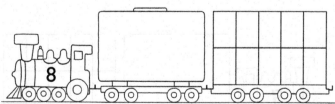

Unit 10

Ten Frames (TRB pp. 58–61)
Number and place value Connect number names, numerals
and quantities, including zero, initially up to 10 and then beyond
(ACMNA002) **AC**

Subitise small collections of objects
(ACMNA003) **AC**

41

Some More, Some Less

✪ Complete the ten frames.

Show 4.

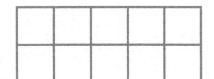

Make 1 less.

Make 1 more.

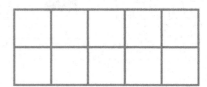

Show seven.

Make 2 less.

Make 2 more.

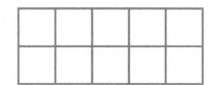

Show nine.

Make two less.

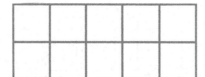

Make 1 more.

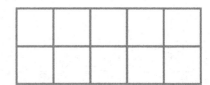

✪ Complete the trains.

42   Unit 10   **Ten Frames** (TRB pp. 58–61)
Number and place value Connect number names, numerals and quantities, including zero, initially up to 10 and then beyond (ACMNA002) AC

Subitise small collections of objects (ACMNA003) AC

STUDENT ASSESSMENT

✿ Use a ten frame to show:

| 3 | 7 | 4 |

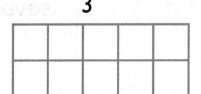 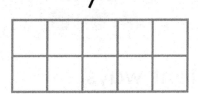 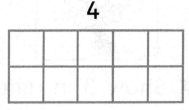

✿ What number can you see?

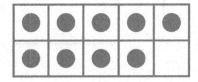 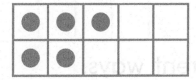 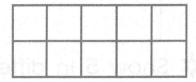

_____ _____ _____

✿ Show two **more** than: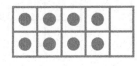

✿ Show one **less** than: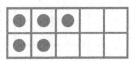

✿ Draw a line to match the numbers.

four

six

10

7

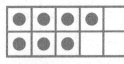

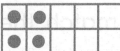

Ten Frames (TRB pp. 58–61)
Number and place value Connect number names, numerals
and quantities, including zero, initially up to 10 and then beyond
(ACMNA002) **AC**

Subitise small collections of objects
(ACMNA003) **AC**

43

Different Ways

✿ Look at the different ways we can show 7.

 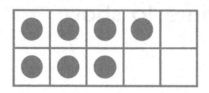

seven

✿ Show 3 in different ways.

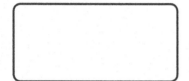 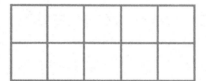 

✿ Show 5 in different ways.

 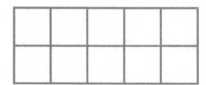 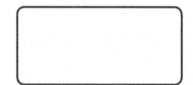

✿ Show 10 in different ways.

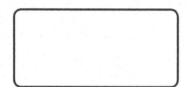 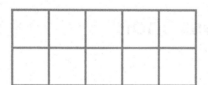 

✿ Show 8 in different ways.

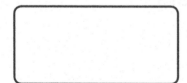 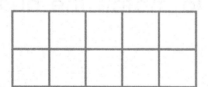 

✿ Draw a line to match the numbers.

  four

6   9   4   10

Lots of Toys

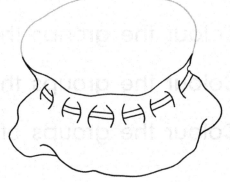

✺ How old are you? _____

✺ Draw that many toys in the bag.

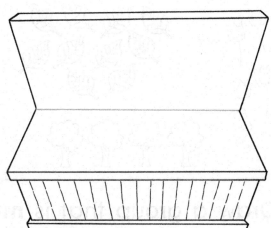

✺ Draw **more** toys than you drew in the bag.

There are _____ toys.

There are _____ books.

Draw **more** books.

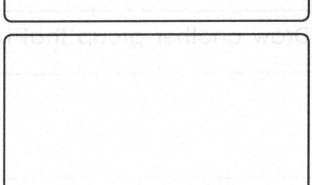

Draw **less** books.

Unit 11 **Counting and Comparing Groups** (TRB pp. 62–65)
Number and place value Connect number names, numerals and quantities, including zero, initially up to 10 and then beyond (ACMNA002) AC

Compare, order and make correspondences between collections, initially to 20, and explain reasoning (ACMNA289) AC

45

Colour the Groups

✪ Colour the groups that are **more than** 5 green.

✪ Colour the groups that are **less than** 5 blue.

✪ Colour the groups of 5 yellow.

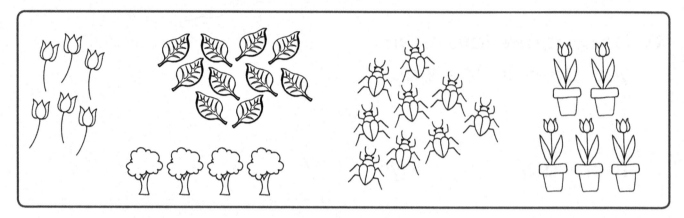

✪ Draw a group that is **more than** 5.

✪ Draw another group that is **more than** 5.

✪ Draw another group that is **more than** 5.

Unit **11** **Counting and Comparing Groups** (TRB pp. 62–65)
Number and place value Connect number names, numerals
and quantities, including zero, initially up to 10 and then beyond
(ACMNA002) AC

Compare, order and make correspondences between
collections, initially to 20, and explain reasoning
(ACMNA289) AC

STUDENT ASSESSMENT

DATE:

✤ Colour the groups that show 3.

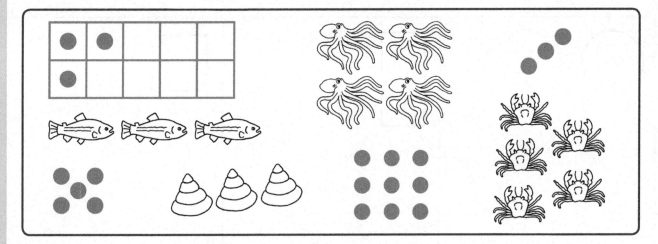

✤ Colour the group that has **less**.

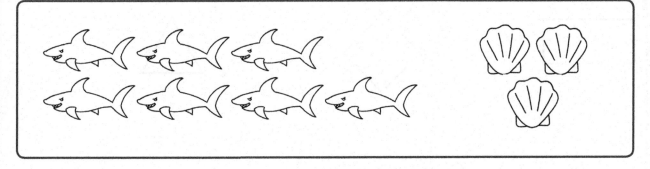

✤ Colour all the groups that are **more than** 5.

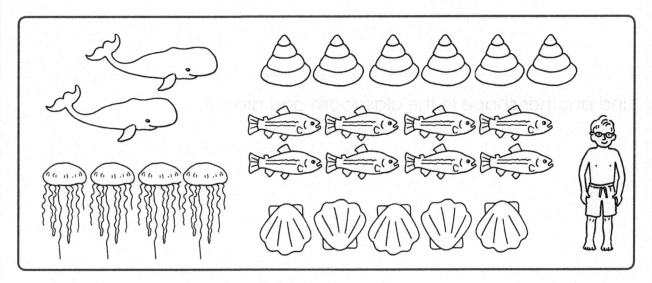

Unit **11**

Counting and Comparing Groups (TRB pp. 62–65)
Number and place value Connect number names, numerals
and quantities, including zero, initially up to 10 and then beyond
(ACMNA002) **AC**

Compare, order and make correspondences between
collections, initially to 20, and explain reasoning
(ACMNA289) **AC**

47

Shapes You Can See

Look at each shape. Draw something from your classroom that has the same shape.

Find another shape in the classroom and draw it.

Unit 12 **2D Shapes** (TRB pp. 66–69)
Shape Sort, describe and name familiar two-dimensional shapes and three-dimensional objects in the environment
(ACMMG009) AC

Find the Shape

☆ Look at the picture below.
Colour all the rectangles blue.

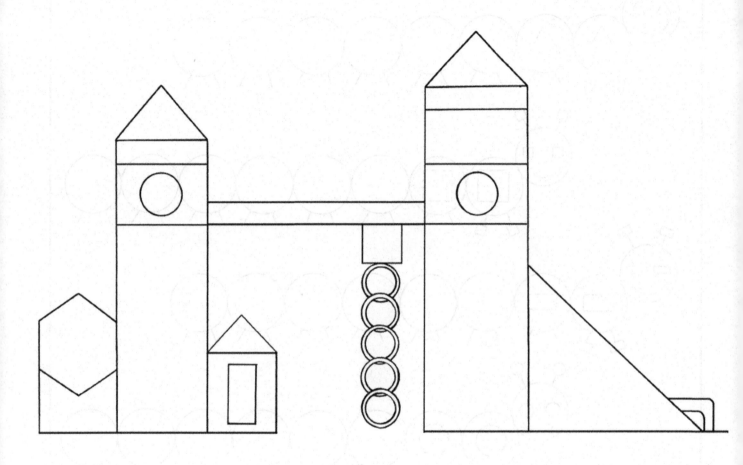

☆ How many rectangles are in the picture?

☆ What other shapes can you see?

_____

_____

Unit 12 **2D Shapes** (TRB pp. 66–69)
Shape Sort, describe and name familiar two-dimensional shapes and three-dimensional objects in the environment
(ACMMG009) **AC**

49

Shape Caterpillars

✿ Finish the patterns on the caterpillars.

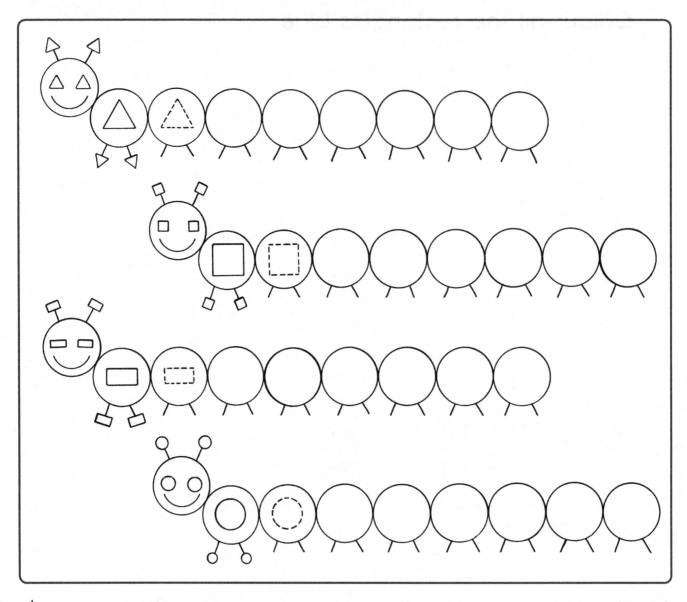

✿ Draw your own shapes on the caterpillar.

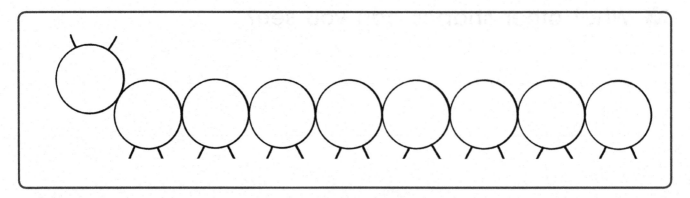

STUDENT ASSESSMENT

✪ Look at all the shapes. Colour the triangles red.

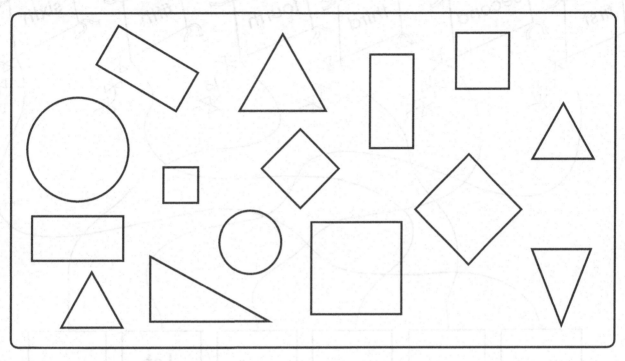

✪ This is a rectangle.

Draw something from your classroom that has a rectangle.

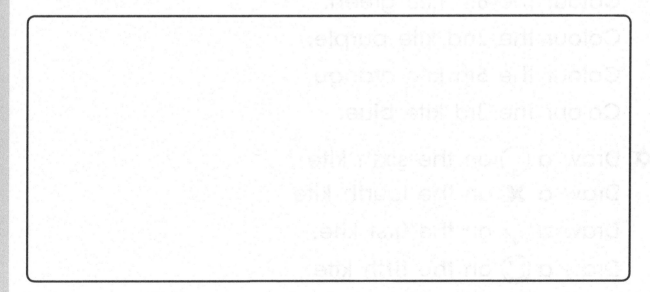

Unit
12

2D Shapes (TRB pp. 66–69)
Shape Sort, describe and name familiar two-dimensional shapes and three-dimensional objects in the environment
(ACMMG009) AC

51

Colourful Kites

✿ Fill in the boxes below.

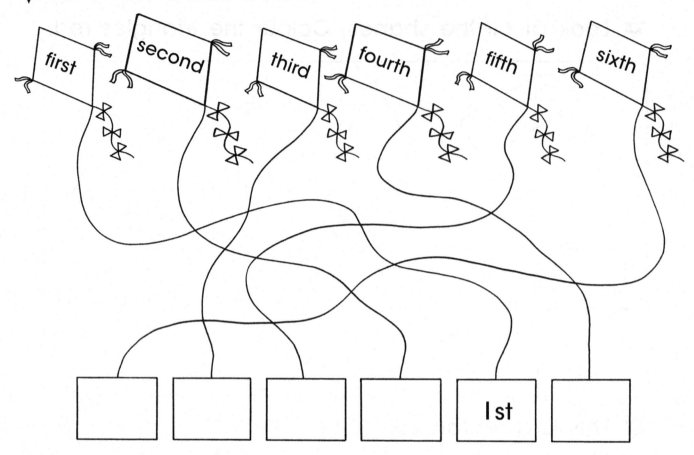

✿ Colour the 1st kite red.

Colour the 4th kite yellow.

Colour the 6th kite green.

Colour the 2nd kite purple.

Colour the 5th kite orange.

Colour the 3rd kite blue.

✿ Draw a ◯ on the sixth kite

Draw a ✖ on the fourth kite.

Draw a ♡ on the first kite.

Draw a ☺ on the fifth kite.

Unit 13 **Ordinal Number** (TRB pp. 70–73)
Number and place value Compare, order and make correspondences between collections,
initially to 20, and explain reasoning
(ACMNA289) **AC**

Clowns in Colour

✪ Colour three hats red.

Colour the other hats any colour you like.

Add more colour to your clowns.

✪ With a partner, take it in turn to describe your clowns.

✪ Now colour in the clowns below to match your partner's clowns.

Your partner will need to tell you which hats to colour red.

They will need to tell you all the other things to colour.

Unit 13 **Ordinal Number** (TRB pp. 70–73)
Number and place value Compare, order and make correspondences between collections, initially to 20, and explain reasoning
(ACMNA289) **AC**

53

Ducks in a Row

☆ Draw a line to match the word with the duck.

One has been done.

fourth   seventh   first   fifth   sixth   third   second

☆ Colour the 3rd and 7th ducks blue.

Colour the 5th and 1st ducks red.

Colour the 4th duck green.

Colour the 6th and 2nd ducks yellow.

54 **Unit 13** **Ordinal Number** (TRB pp. 70–73)
Number and place value Compare, order and make correspondences between collections,
initially to 20, and explain reasoning
(ACMNA289) **AC**

✿ Draw a ✖ on the **first** car.

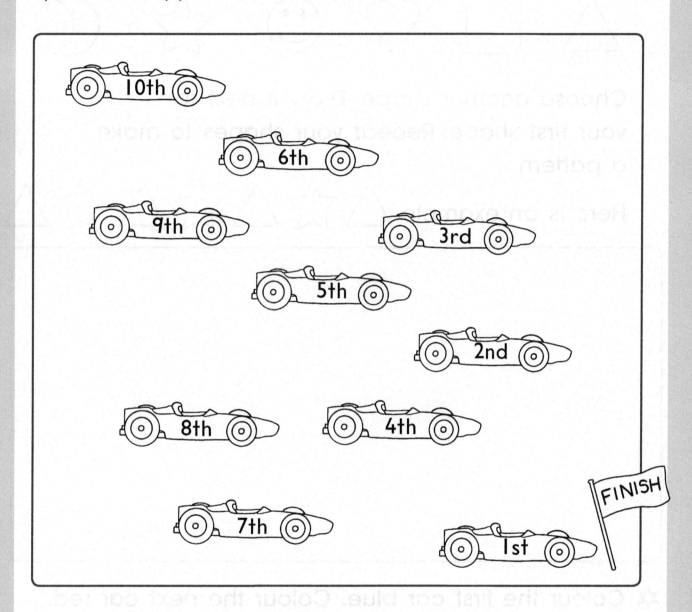

✿ Colour the **fourth** car red.

Colour the **seventh** car green.

Colour the **second** car yellow.

Colour the **fifth** car blue.

Unit
13 **Ordinal Number** (TRB pp. 70–73)
Number and place value Compare, order and make correspondences between collections,
initially to 20, and explain reasoning
(ACMNA289) **AC**

55

Make a Pattern

✿ **Choose one shape and draw it in the box.**

Choose another shape. Draw it next to your first shape. Repeat your shapes to make a pattern.

Here is an example:

✿ **Colour the first car blue. Colour the next car red. Repeat your colouring to make a pattern.**

✿ **Draw your own pattern on another sheet of paper.**

Patterns (TRB pp. 74–77)
Patterns and algebra Sort and classify familiar objects and explain the basis for these classifications.
Copy, continue and create patterns with objects and drawings
(ACMNA005) **AC**

Musical Patterns

✿ Look at the pattern. Make the pattern by clapping.

Draw the same pattern.

✿ Make another pattern using  and 🪇 .

Clap this pattern.

✿ Finish the pattern below.

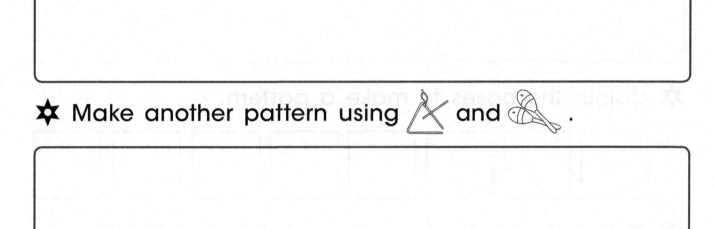

✿ Make a musical pattern using some of the instruments above.

Unit 14 **Patterns** (TRB pp. 74–77)
Patterns and algebra Sort and classify familiar objects and explain the basis for these classifications.
Copy, continue and create patterns with objects and drawings
(ACMNA005) **AC**

57

Shapes and Patterns

✿ Copy the pattern.

✿ Colour the boxes to make a pattern.

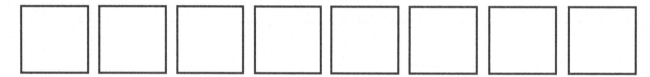

✿ Colour the circles to make a pattern.

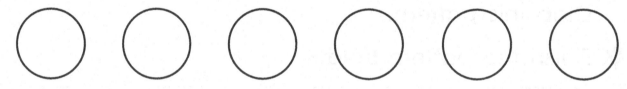

✿ Finish the patterns.

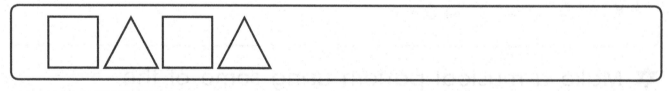

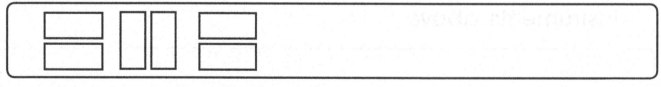

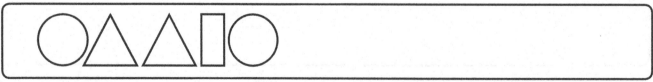

Patterns (TRB pp. 74–77)
Patterns and algebra Sort and classify familiar objects and explain the basis for these classifications.
Copy, continue and create patterns with objects and drawings
(ACMNA005) **AC**

STUDENT ASSESSMENT

✿ **Colour to make a pattern.**

✿ **Copy the pattern.**

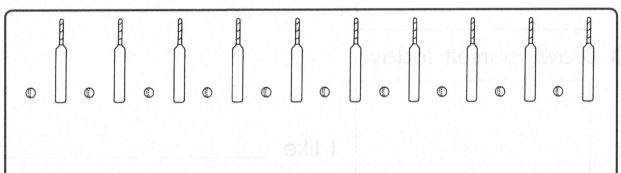

✿ **Finish the patterns.**

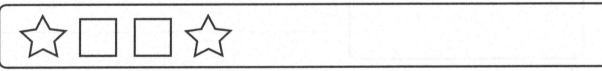

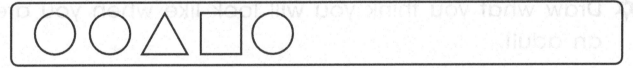

✿ **Draw your own pattern.**

Unit
14

Patterns (TRB pp. 74–77)
Patterns and algebra Sort and classify familiar objects and explain the basis for these classifications.
Copy, continue and create patterns with objects and drawings
(ACMNA005) **AC**

59

My Timeline

✦ Draw yourself when you were a baby.

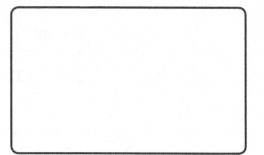

I liked _____

_____

✦ Draw yourself today.

I like _____

_____

✦ Draw what you think you will look like in high school.

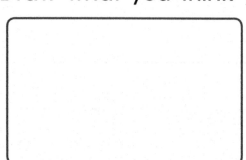

I think I will be _____

_____

✦ Draw what you think you will look like when you are an adult.

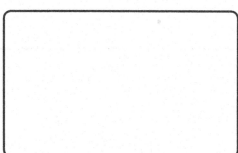

I think I will be _____

_____

My School Day

I wake up.

Then I _____

_____

_____

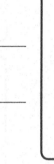

I come to school by

_____

_____

At school, I like to

_____

_____

Unit 15 **Time** (TRB pp. 78–81)
Using units of measurement Compare and order the duration of events using the everyday language of time
(ACMMG007) **AC**

61

Yesterday

✸ **Think about yesterday.**

Draw something that
you liked doing
in the morning.

✸ Draw something that
you liked doing
in the afternoon.

✸ Draw something that
you liked doing
in the evening.

✸ **What was your favourite time of the day?**

_____

Why? _____

Unit **15**

STUDENT ASSESSMENT

✿ Look at the pictures. There is a 1 under the first thing the boy does.

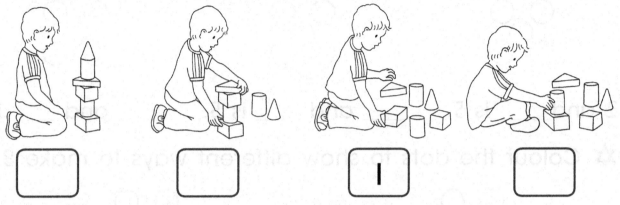

| | | 1 | |

Write 2 under what he does next.

Write 3 under what he does next.

Write 4 under the last thing he does.

✿ Draw something you do:

- after school.

- before school.

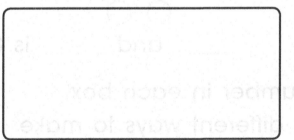

✿ Draw something you do at these times:

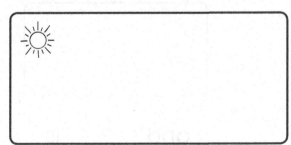

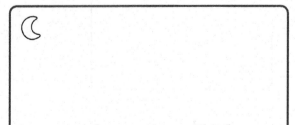

Unit **15** **Time** (TRB pp. 78–81)
Using units of measurement Compare and order the duration of events using the everyday language of time
(ACMMG007) **AC**

63

Colour the Dots

✪ Colour 2 dots blue and the others red.

✪ Colour the dots to show different ways to make 5.

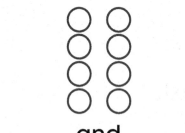

2 and ____ is 5.   ____ and ____ is 5.   ____ and ____ is 5.

✪ Colour the dots to show different ways to make 8.

_____ and _____ is 8.   _____ and _____ is 8.

_____ and _____ is 8.   _____ and _____ is 8.

✪ Draw dots for another number in each box.
Colour the dots to show different ways to make
the number.

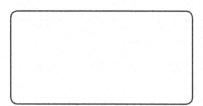   

_____ and _____ is _____   _____ and _____ is _____

 Understanding Numbers to 10 (TRB pp. 82–85)
Number and place value Connect number names, numerals and quantities, including zero,
initially up to 10 and then beyond
(ACMNA002) **AC**

Lovely Ladybirds

� Look at the ladybird. Write about her dots.

_____ and _____ is _____

✶ Draw different dots on each side of the ladybirds to show the same number.

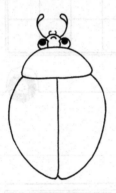   

_____ and _____ is _____   _____ and _____ is _____

✶ Show different ways to make 9.

Write about what you have drawn.

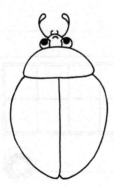   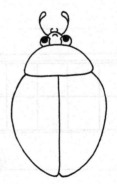

_____   _____

Unit 16 **Understanding Numbers to 10** (TRB pp. 82–85)
Number and place value Connect number names, numerals and quantities, including zero, initially up to 10 and then beyond
(ACMNA002) **AC**

65

Ten Bus

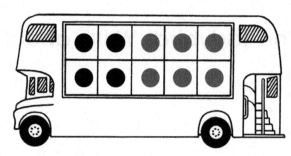

4 and 6 is 10.

✸ Draw red dots and blue dots on each bus to show different ways to make 10.

✸ Write a number story for each bus.

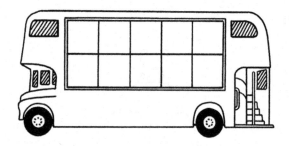

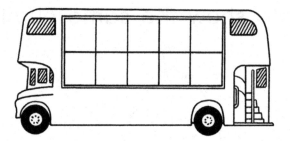

_____

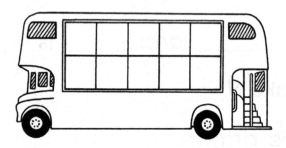

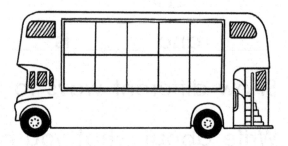

_____

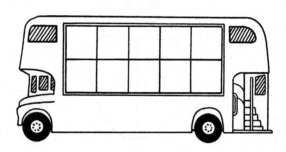

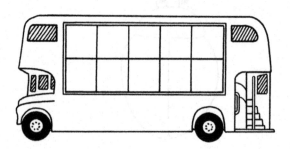

_____

66 Unit 16 **Understanding Numbers to 10** (TRB pp. 82–85)
Number and place value Connect number names, numerals and quantities, including zero,
initially up to 10 and then beyond
(ACMNA002) AC

STUDENT ASSESSMENT

✿ There are 6 puppies in a litter. Some are brown and some are black. Colour the puppies to show what the litter might look like.

✿ How else might the litter look?

✿ 10 balls are in a box. Some are blue and some are red. Colour the balls to show one way they might look.

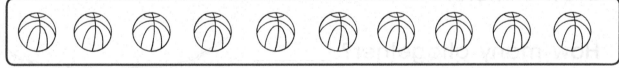

✿ There are 7 eggs in a nest. Some are white and some are brown. Draw one way they might look.

Unit
16

Understanding Numbers to 10 (TRB pp. 82–85)
Number and place value Connect number names, numerals and quantities, including zero, initially up to 10 and then beyond
(ACMNA002) AC

67

Lots of Pets

✿ There are 4 fish.
Draw 3 more.

How many altogether? _____

✿ There are 6 puppies.
Draw 2 more.

How many altogether? _____

✿ There are 2 ducks.
Draw 5 more.

How many altogether? _____

✿ There is 1 kitten.
Draw 3 more.

How many altogether? _____

Unit 17 **Beginning Addition** (TRB pp. 86–89)
Number and place value Represent practical situations to model addition and sharing
(ACMNA004) AC

In the Paddocks

✡ Draw 3 sheep and draw 3 sheep.

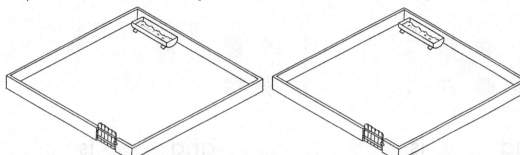

How many altogether?

_____

✡ Draw 3 cows and draw 6 cows.

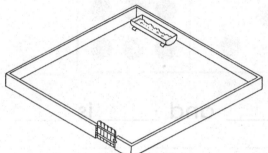
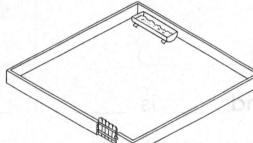

How many altogether?

_____

✡ Draw 4 horses and draw 3 horses.

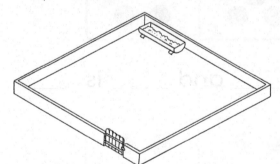
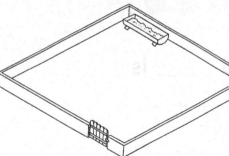

How many altogether?

_____

✡ Write your own story.

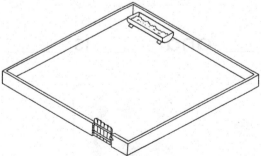
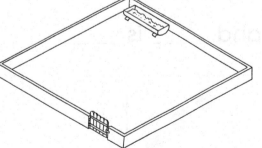

There are _____ and _____ .

There are _____ altogether.

Unit **17** **Beginning Addition** (TRB pp. 86–89)
Number and place value Represent practical situations to model addition and sharing
(ACMNA004) *AC*

69

Dots on Dominoes

Fill in the numbers.

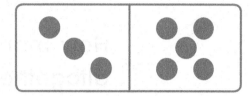

_____ and _____ is _____

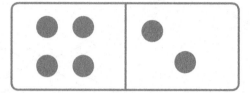

_____ and _____ is _____

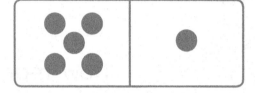

_____ and _____ is _____

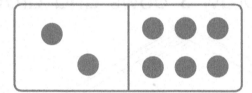

_____ and _____ is _____

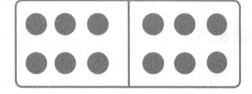

_____ and _____ is _____

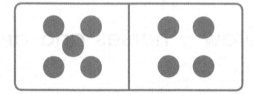

_____ and _____ is _____

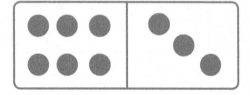

_____ and _____ is _____

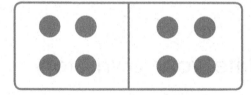

_____ and _____ is _____

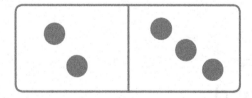

_____ and _____ is _____

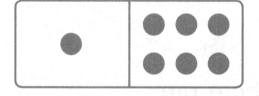

_____ and _____ is _____

Beginning Addition (TRB pp. 86–89)
Number and place value Represent practical situations to model addition and sharing
(ACMNA004) **AC**

STUDENT ASSESSMENT

There are _____ books and _____ books.

There are _____ books altogether.

✪ Draw 5 more pencils.

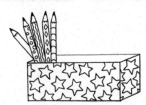

There are _____ pencils altogether.

✪ Look at the shells. Write a story about them.

_____

✪ Look at the domino.

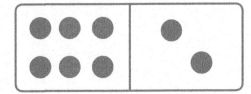

How many dots altogether? _____

How did you work that out? _____

_____

Unit
17
Beginning Addition (TRB pp. 86–89)
Number and place value Represent practical situations to model addition and sharing
(ACMNA004) **AC**

71

Sorting Shapes

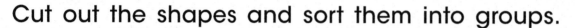

You will need:

- a copy of **BLM 41 'Shape of Things'**
- scissors, glue, paper

Cut out the shapes and sort them into groups.

Choose one group of shapes to paste below.

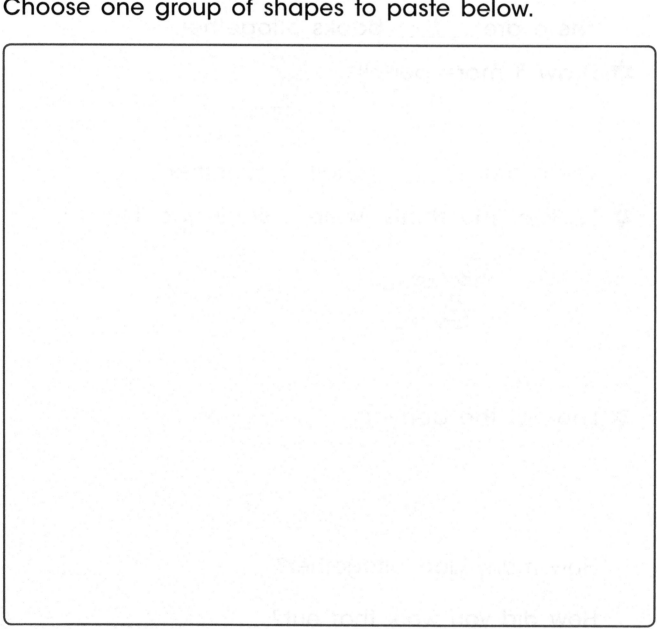

Paste your other groups on another sheet of paper.

Unit 18 **More About Shapes and Objects** (TRB pp. 90–93)
Shape Sort, describe and name familiar two-dimensional shapes and three-dimensional objects in the environment
(ACMMG009) **AC**

What Could It Be?

DATE:

Look around. Draw things that are the same as these shapes.

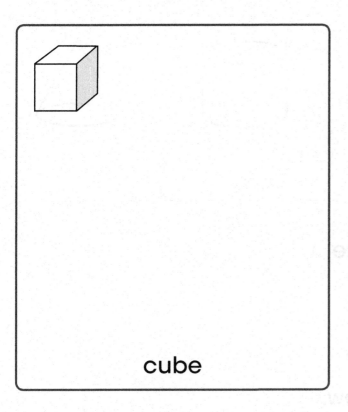

cube

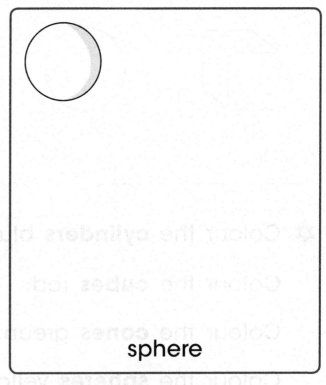

sphere

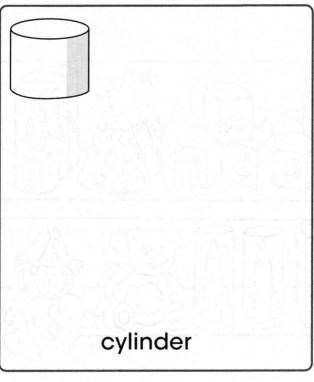

cylinder

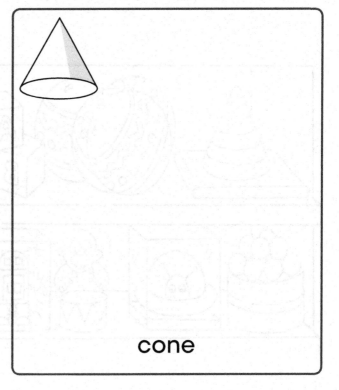

cone

Unit 18 **More About Shapes and Objects** (TRB pp. 90–93)
Shape Sort, describe and name familiar two-dimensional shapes and three-dimensional objects in the environment
(ACMMG009) **AC**

73

Can You Find the Shapes?

✿ Write the name of each shape.

_____ _____ _____ _____

✿ Colour the **cylinders** blue.

Colour the **cubes** red.

Colour the **cones** green.

Colour the **spheres** yellow.

Unit 18 **More About Shapes and Objects** (TRB pp. 90–93)
Shape Sort, describe and name familiar two-dimensional shapes and three-dimensional objects in the environment
(ACMMG009) **AC**

STUDENT ASSESSMENT

✿ Write the name of each shape.

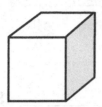

_____

_____

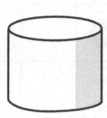

_____

_____

✿ Draw something that is the same as
each shape.

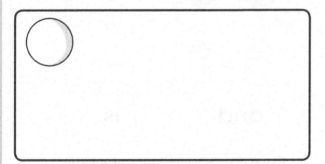

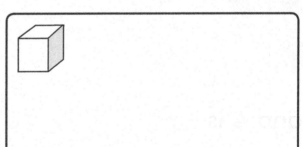

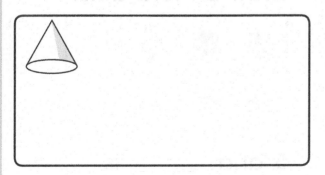

Unit
18
More About Shapes and Objects (TRB pp. 90–93)
Shape Sort, describe and name familiar two-dimensional shapes and three-dimensional objects in the environment
(ACMMG009) AC

75

Addition with Ten Frames

Draw 5 more dots.

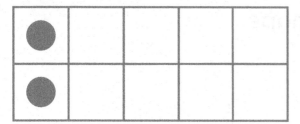

2 and 5 is _____

Draw 2 more dots.

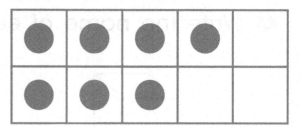

7 and _____ is _____

Draw 2 more dots.

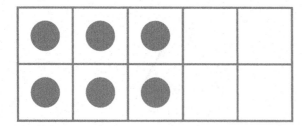

6 and 2 is _____

Draw 5 more dots.

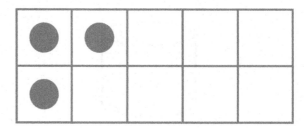

3 and _____ is _____

Draw 4 more dots.

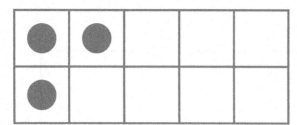

3 and 4 is _____

Draw 1 more dot.

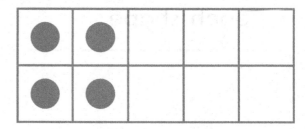

4 and _____ is _____

Draw 9 more dots.

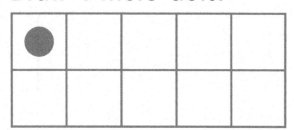

1 and 9 is _____

Draw 2 more dots.

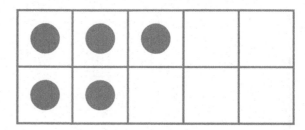

5 and _____ is _____

Dominoes and Dice

✸ Draw 4 and **one** more.

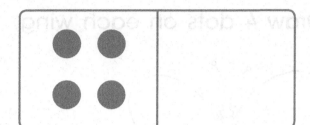

Draw 1 and **one** more.

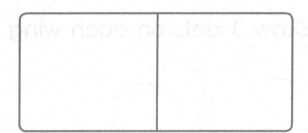

Draw 6 and **one** more.

Draw 5 and **one** more.

Draw 3 and **one** more.

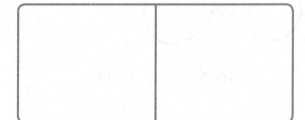

Draw 2 and **one** more.

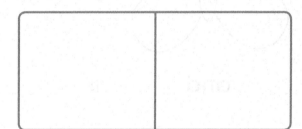

✸ Draw **one** more than:

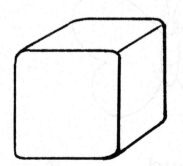

Draw **one** more than:

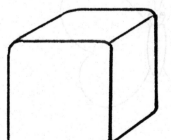

Unit 19 **More About Addition** (TRB pp. 94–97)
Number and place value Represent practical situations to model addition and sharing
(ACMNA004) **AC**

77

Beautiful Butterflies

Butterflies have the same pattern on both wings.

Draw 3 dots on each wing.   Draw 4 dots on each wing.

3 and 3 is _____   4 and 4 is _____

Draw 4 dots on each wing.   Draw 2 dots on each wing.

_____ and _____ is _____   _____ and _____ is _____

Draw dots on these butterflies.   Write a number sentence.

_____ and _____ is _____   _____ and _____ is _____

Unit 19   **More About Addition** (TRB pp. 94–97)
Number and place value Represent practical situations to model addition and sharing
(ACMNA004) **AC**

Unit 19 STUDENT ASSESSMENT

✸ **Use the ten frames to work out the answers.**

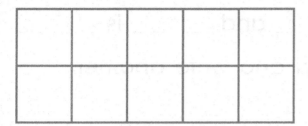

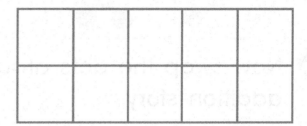

4 and 5 is _____   2 and 8 is _____

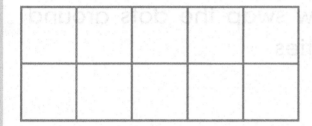

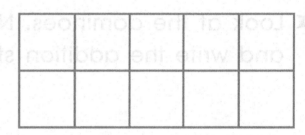

6 and 3 is _____   3 and 2 is _____

✸ I more than 7 is _____   Double 3 is _____

  I more than 4 is _____   Double I is _____

  I more than 3 is _____   Double 5 is _____

  I more than 2 is _____   Double 6 is _____

  I more than 8 is _____   Double 2 is _____

  I more than 6 is _____   Double 4 is _____

Unit 19 **More About Addition** (TRB pp. 94–97)
Number and place value Represent practical situations to model addition and sharing
(ACMNA004) AC

79

Swap It Around

✦ Look at the domino and write an addition story.

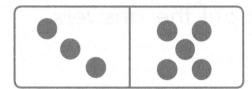 _____ and _____ is _____

✦ Now swap the dots around and write another addition story.

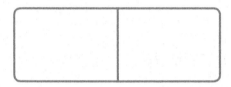 _____ and _____ is _____

✦ Look at the dominoes. Now swap the dots around and write the addition stories.

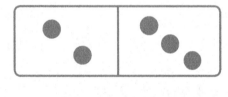   

_____   _____

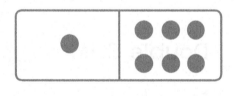   

_____   _____

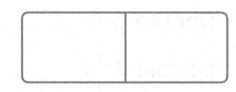   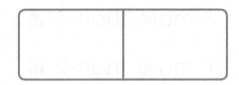

_____   _____

✦ Write what you learned. _____

_____

Moving Ahead with Addition (TRB pp. 98–101)
Number and place value Represent practical situations to model addition and sharing
(ACMNA004) **AC**

Jumping Kangaroos

✸ If a 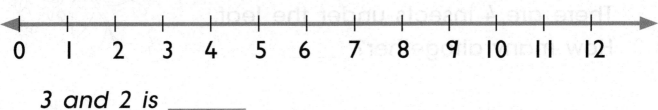 was on 3 and jumped forwards 2 places, what number would it be on?

0 1 2 3 4 5 6 7 8 9 10 11 12

3 and 2 is _____

✸ If a was on 1 and it jumped 6 places, what number would it be on?

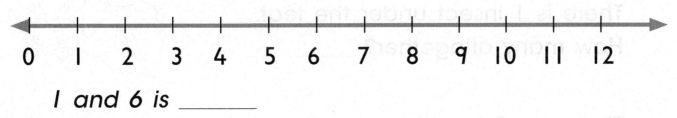

0 1 2 3 4 5 6 7 8 9 10 11 12

1 and 6 is _____

✸ If a was on 7 and it jumped 3 places, what number would it be on?

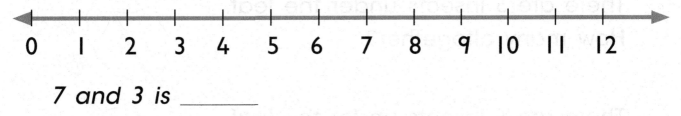

0 1 2 3 4 5 6 7 8 9 10 11 12

7 and 3 is _____

✸ If a was on 4 and it jumped 4 places, what number would it be on?

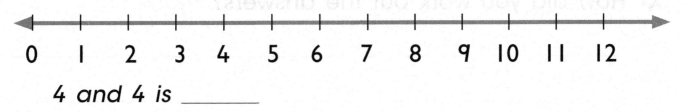

0 1 2 3 4 5 6 7 8 9 10 11 12

4 and 4 is _____

Unit 20
Moving Ahead with Addition (TRB pp. 98–101)
Number and place value Represent practical situations to model addition and sharing
(ACMNA004) AC

81

Insects! Insects!

✦ Some insects are hiding!

There are 4 insects under the leaf.
How many altogether? _____

There are 3 insects under the leaf.
How many altogether? _____

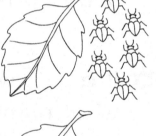

There is 1 insect under the leaf.
How many altogether? _____

There are 2 insects under the leaf.
How many altogether? _____

There are 5 insects under the leaf.
How many altogether? _____

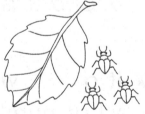

There are 6 insects under the leaf.
How many altogether? _____

✦ How did you work out the answers?

_____

_____

82 | Unit 20 | **Moving Ahead with Addition** (TRB pp. 98–101)
Number and place value Represent practical situations to model addition and sharing
(ACMNA004) AC

STUDENT ASSESSMENT

✧ 4 and 5 is 9.

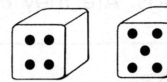

Use 4 and 5 to show another way to make 9.

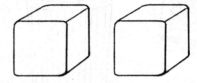

✧ 3 and 4 is 7.

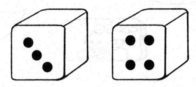

Use 3 and 4 to show another way to make 7.

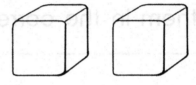

✧ Use the number line to work out:

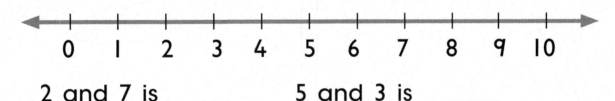

2 and 7 is _____   5 and 3 is _____

✧ There are 4 marbles in my closed hand. How many altogether? _____

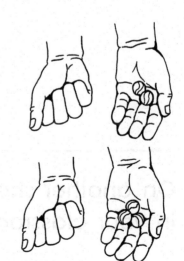

There are 6 marbles in my closed hand. How many altogether? _____

Unit 20 **Moving Ahead with Addition** (TRB pp. 98–101)
Number and place value Represent practical situations to model addition and sharing
(ACMNA004) **AC**

83

Is It Heavy or Light?

✯ Pick up these things. Are they **heavy** or **light**?

✯ Draw them in the correct box.

Heavy	Light

✯ On another sheet of paper, draw some other things in your classroom that are **heavy** or **light**.

Unit 21 **Mass** (TRB pp. 102–105)
Using units of measurement Use direct and indirect comparisons to decide which is longer, heavier or holds more, and explain reasoning in everyday language
(ACMMG006) AC

Heavier Than a Book

✵ Pick up a book.

Now find 5 things that are **heavier** than your book. Draw them.

✵ What was the **heaviest** thing you found?

✵ What was the **lightest** thing you found?

Unit 21 **Mass** (TRB pp. 102–105)
Using units of measurement Use direct and indirect comparisons to decide which is longer, heavier or holds more, and explain reasoning in everyday language
(ACMMG006) AC

85

Sorting Toys

✪ Draw a toy in the top box.

Draw a toy that is **heavier** than your toy.

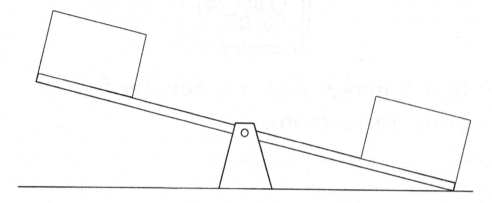

✪ Draw your toy in the bottom box.

Draw a toy that is **lighter** than your toy.

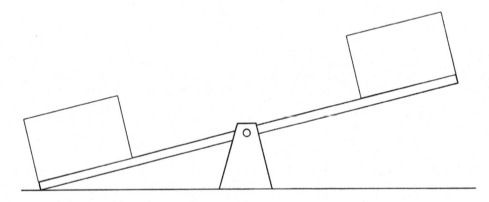

✪ Draw all your toys from heaviest to lightest.

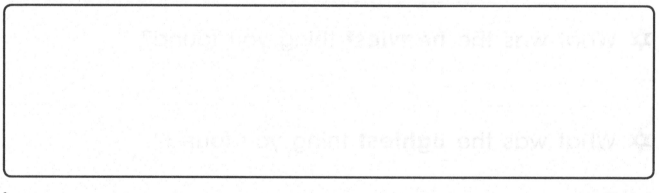

heavy ◀──────────────────────▶ light

Unit 21
Mass (TRB pp. 102–105)
Using units of measurement Use direct and indirect comparisons to decide which is longer, heavier or holds more, and explain reasoning in everyday language
(ACMMG006) **AC**

✿ Draw something that is **heavy**.

✿ Draw something that is **light**.

✿ Draw something that is **heavier** than a 🐻.

✿ Colour the **lightest** toy.

Unit **21** **Mass** (TRB pp. 102–105)
Using units of measurement Use direct and indirect comparisons to decide which is longer,
heavier or holds more, and explain reasoning in everyday language
(ACMMG006) AC

87

Number Search

✪ Colour the numbers.

11 is red, 12 is green, 13 is purple, 14 is light blue,
15 is dark blue, 16 is orange, 17 is yellow,
18 is brown, 19 is pink and 20 is grey.

11	12	13	14	15	16	17
18	19	20	16	18	17	19
20	14	11	13	15		19
18	17	17	15	14	13	
12	11	19	17	15	13	11
18	16	14	12	13	11	16
11	19	14	17	11	12	15	18
11	11	15	12	16	13	17
14	19	18	12	13	14
15	16	17	18	19	14	17

✪ Write how many of each number you found.

11	12	13	14	15	16	17	18	19	20

88 Unit 22 **Numbers Beyond 10** (TRB pp. 106–109)
Number and place value Establish understanding of the language and processes of counting
by naming numbers in sequences, initially to and from 20, moving from any starting point
(ACMNA001) AC

Building Climb

✿ Write a number on each floor of this building. The first two have been done for you.

✿ Play this game with a partner to see who can climb their building first.

You will need:

- a dice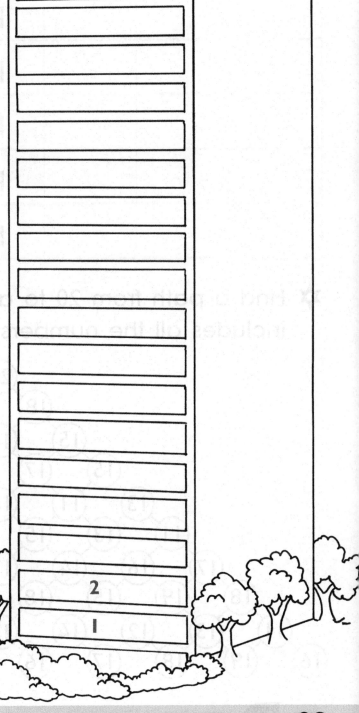
- counters

How to play:

- Roll a dice 4 times and move that many spaces.

- Now your partner should do the same.

- If you have made it to the top, score 1 point. Otherwise the person on the highest floor scores 1 point.

- The first person with 10 points wins.

2

1

Unit **22** **Numbers Beyond 10** (TRB pp. 106–109)
Number and place value Establish understanding of the language and processes of counting by naming numbers in sequences, initially to and from 20, moving from any starting point
(ACMNA001) **AC**

89

Number Challenge

✿ Complete the counting sequence in each row.

12	13	14	15	16	17	18
			17			
			12			
			16			
			14			
			13			
			18			

✿ Find a path from 20 to a number at the bottom that includes all the numbers from 11 to 20.

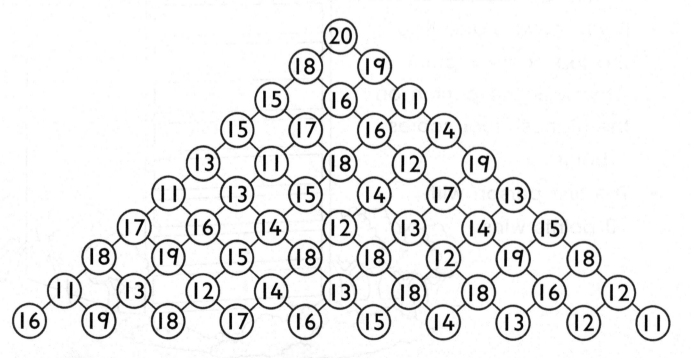

Unit 22

Numbers Beyond 10 (TRB pp. 106–109)
Number and place value Establish understanding of the language and processes of counting
by naming numbers in sequences, initially to and from 20, moving from any starting point
(ACMNA001) **AC**

STUDENT ASSESSMENT

DATE:

✪ Fill in the missing numbers.

1	2	3			6		8		10
11			14			17			20

✪ Write the numbers in the correct order.

17   15   18   16

_____

12   14   13   11

_____

10   12   9   11

_____

20   18   19   17

_____

✪ Write the five numbers that come
after 10 when you are counting.

_____

Unit
22   **Numbers Beyond 10** (TRB pp. 106–109)
Number and place value Establish understanding of the language and processes of counting
by naming numbers in sequences, initially to and from 20, moving from any starting point
(ACMNA001) AC

91

Counting Back

You will need:

a copy of **BLM 47 'Number Tiles 11–20',** scissors and glue

✪ Cut out the cards and shuffle them.
Place the cards face down.

Turn over a card. Write the number in the box below.
Now count as far back as you can.

Repeat for all the other boxes.

✪ Turn over 2 cards. Write them in the boxes.

Write the missing numbers in between.

✪ Try this again.

Counting Again (TRB pp. 110–113)
Number and place value Establish understanding of the language and processes of counting
by naming numbers in sequences, initially to and from 20, moving from any starting point
(ACMNA001) **AC**

Number Lines

☼ Write the numbers from 1 to 20 on the number line.

☼ Look at the number line. Find a number that is **more than**:

6 _____   15 _____   18 _____

17 _____   12 _____   10 _____

☼ Look at the number line. Find a number that is **less than**:

11 _____   19 _____   20 _____

14 _____   16 _____   13 _____

☼ I ate 14 bananas. My friend ate **more** bananas than me. Circle the numbers to show how many bananas she might have eaten.

17   12   14   9   10   15   18

Unit 23 **Counting Again** (TRB pp. 110–113)
Number and place value Connect number names, numerals and quantities, including zero, initially up to 10 and then beyond
(ACMNA002) **AC**

93

Comparing Groups

✦ Write how many blocks are in each group.

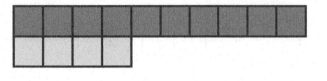
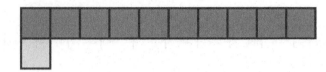

_____   _____

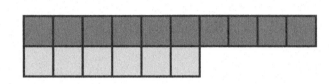
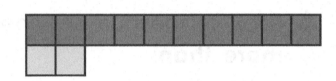

_____   _____

✦ Draw blocks to match each number.

19	17

✦ Draw blocks to show a number that is **more than**:

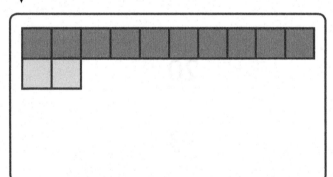
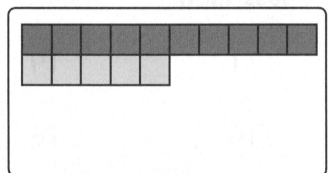

✦ Draw blocks to show a number that is **less than**:

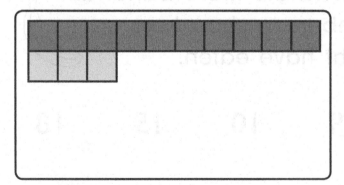
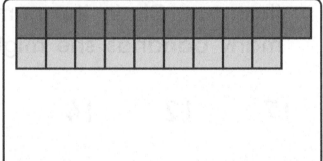

Unit 23 **Counting Again** (TRB pp. 110–113)
Number and place value Compare, order and make correspondences between collections,
initially to 20, and explain reasoning
(ACMNA289) AC

Unit **23** STUDENT ASSESSMENT

✪ Look at the number in the box. Count on as far as you can.

6 _____

11 _____

10 _____

✪ Look at the number in the box. Count backwards as far as you can.

9 _____

15 _____

13 _____

✪ Fill in the missing numbers.

9 _____ 15

19 _____ 11

12 _____ 17

✪ Draw blocks to match the number.

15

✪ Draw blocks to show a group that is **more than** 13.

Unit **23** **Counting Again** (TRB pp. 110–113)
Number and place value Establish understanding of the
language and processes of counting by naming numbers in
sequences, initially to and from 20, moving from any starting point
(ACMNA001) AC

Connect number names,
numerals and quantities,
including zero, initially up to
10 and then beyond
(ACMNA002) AC

Compare, order and make
correspondences between
collections, initially to 20, and
explain reasoning
(ACMNA289) AC

95

Numbers on Ten Frames

✦ Write how many.

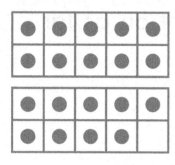

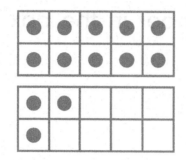

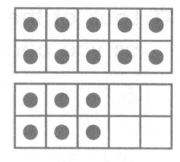

_____

_____

_____

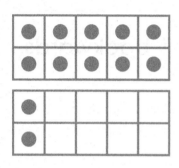

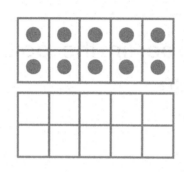

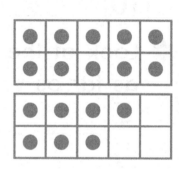

_____

_____

_____

✦ Show this many.

14

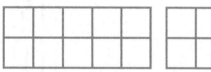

18

12

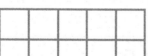

15

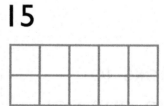

20

11

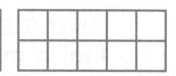

 Unit 24 **More About Numbers to 20** (TRB pp. 114–117)
Number and place value Connect number names, numerals
and quantities, including zero, initially up to 10 and then beyond
(ACMNA002) **AC**

Compare, order and make correspondences between
collections, initially to 20, and explain reasoning
(ACMNA289) **AC**

Numerals, Words and Pictures

Write or draw the numeral, word or picture.

11	eleven	(tulips)
	twelve	
	thirteen	
		(pencils)
	fifteen	
		(tennis balls)
17		
18		
		(bubbles)
	twenty	

Unit 24 **More About Numbers to 20** (TRB pp. 114–117)
Number and place value Connect number names, numerals
and quantities, including zero, initially up to 10 and then beyond
(ACMNA002) **AC**

Compare, order and make correspondences between
collections, initially to 20, and explain reasoning
(ACMNA289) **AC**

97

Showing the Same Number

Your teacher will give you a card from
BLM 1 'Number Cards 0–5'.

Paste your number card in the middle.

Fill in the spaces around it.

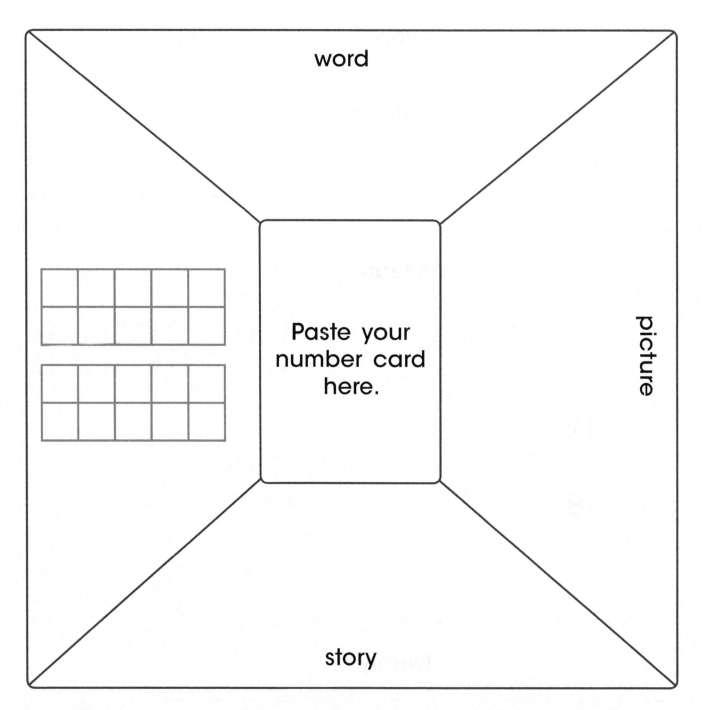

word

picture

Paste your
number card
here.

story

More About Numbers to 20 (TRB pp. 114–117)
Number and place value Connect number names, numerals
and quantities, including zero, initially up to 10 and then beyond
(ACMNA002) **AC**

Compare, order and make correspondences between
collections, initially to 20, and explain reasoning
(ACMNA289) **AC**

Unit 24 STUDENT ASSESSMENT

✿ Write how many.

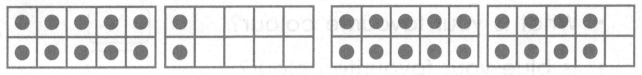

_____   _____   _____

✿ Show the numbers on ten frames.

17   13

✿ Write the matching word for the number.

12 _____   17 _____

15 _____   11 _____

✿ Write the matching number.

thirteen _____   eighteen _____

sixteen _____   fourteen _____

✿ Draw lines matching pictures, words and numbers.

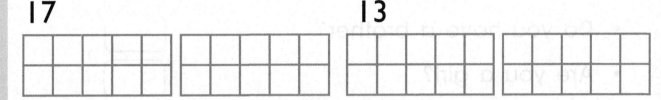

✔✔✔✔✔✔✔✔✔✔   ten   10

🌼🌼🌼🌼🌼🌼🌼🌼🌼🌼🌼🌼🌼🌼🌼🌼   sixteen   11

★★★★★★★★★★★★★★★★★★   eleven   13

●●●●●●●●●●●●●   thirteen   16

Unit 24

More About Numbers to 20 (TRB pp. 114–117)
Number and place value Connect number names, numerals
and quantities, including zero, initially up to 10 and then beyond
(ACMNA002) AC

Compare, order and make correspondences between
collections, initially to 20, and explain reasoning
(ACMNA289) AC

99

"Yes" and "No" Questions

✭ Read the questions. Put a ✔ if you can answer "yes" or "no".

- What is your favourite colour? ☐
- Is blue your favourite colour? ☐
- How do you come to school? ☐
- Do you walk to school? ☐
- Do you have a brother? ☐
- Are you a girl? ☐
- What do you have for lunch? ☐
- Did you have fruit for lunch today? ☐

✭ Write your own "yes" or "no" question.

_____

_____

✭ Choose one of the "yes" or "no" questions from the list. Write it here.

_____

_____

✭ Ask two of your friends the question. Write what you find out.

_____

Questions

✪ Write two "yes" or "no" questions about these children.

_____

_____

✪ Choose one of the questions. Show how you could present the answers.

Unit 25 **Our Class** (TRB pp. 118–121)
Data representation and interpretation Answer yes/no questions to collect information
(ACMSP011) AC

101

Sandwiches for Lunch

✸ Look at the display.

Do you have a sandwich for lunch?

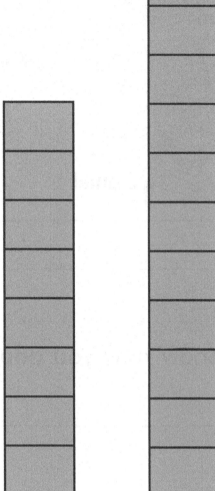

Yes   No

✸ Do most children have a sandwich for lunch?

_____

✸ How many children have a sandwich for lunch?

_____

✸ How many children have something else for lunch?

_____

✸ If you asked your class this question, would the display look the same?

_____

✸ Why do you think that?

_____

_____

Unit 25
STUDENT ASSESSMENT

✦ ✔ the questions you can answer with "yes" or "no".

- Are your socks white?

- What colour are your socks?

- Have you been to the beach?

- Do you go to school?

- How old are you?

✦ Some children were asked "Do you have a purple pencil?"

They said:

On another sheet of paper, draw a display to show the children's answers.

Write what the display tells us.

_____

_____

Unit 25 **Our Class** (TRB pp. 118–121)
Data representation and interpretation Answer yes/no questions to collect information
(ACMSP011) AC

103

Full

✸ Circle the containers that are **full**.

✸ Work with a partner.
Find out how many pencils
a pencil case holds.

Write what you found out.

_____

_____

_____

How Much Does It Hold? (TRB pp. 122–125)
Using units of measurement Use direct and indirect comparisons to decide which is longer,
heavier or holds more, and explain reasoning in everyday language
(ACMMG006) **AC**

Which Container Holds More?

✭ Colour the container that holds **more**.

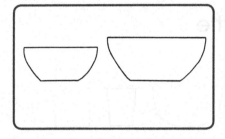 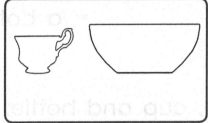 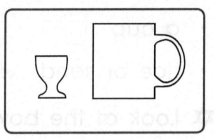

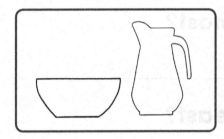 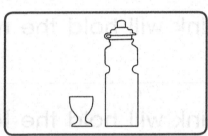 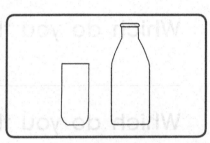

✭ How many cubes will fit into your shoe?
Have a guess first. Then check by filling
your shoe with cubes.

I guess _____ cubes.

My shoe holds _____ cubes.

✭ Find something in the
classroom that holds
more cubes than
your shoe. Draw it.

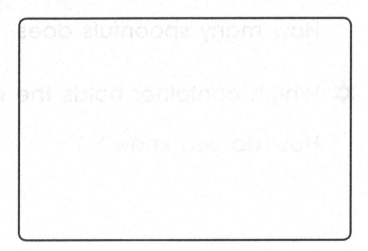

Unit 26

How Much Does It Hold? (TRB pp. 122–125)
Using units of measurement Use direct and indirect comparisons to decide which is longer,
heavier or holds more, and explain reasoning in everyday language
(ACMMG006) AC

105

How Many Spoonfuls?

You will need:

- a spoon
- a cup
- rice or sand

- a bowl
- a bottle

✪ Look at the bowl, cup and bottle.

Which do you think will hold the **most**?

_____

Which do you think will hold the **least**?

_____

✪ Use the spoon and the rice (or sand) to find out how much each container holds.

How many spoonfuls does ⊔ hold? _____

How many spoonfuls does ⌣ hold? _____

How many spoonfuls does 🍼 hold? _____

✪ Which container holds the **most**? _____

How do you know?

_____

_____

How Much Does It Hold? (TRB pp. 122–125)
Using units of measurement Use direct and indirect comparisons to decide which is longer, heavier or holds more, and explain reasoning in everyday language
(ACMMG006) **AC**

DATE:

You will need:

- a paper or plastic cup
- counters

✿ Circle the containers that are **empty**.

  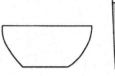

✿ Colour the container that holds **more**.

 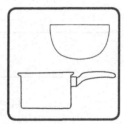  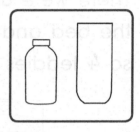

✿ Find something in the classroom that holds **more than** the paper cup.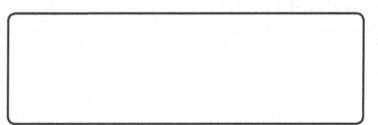

Draw it.

How do you know that it holds **more**?

_____

✿ Guess how many counters will fit into the cup. Check by filling it with counters.

I guess _____ counters.

The cup holds _____ counters.

Unit **26** **How Much Does It Hold?** (TRB pp. 122–125)
Using units of measurement Use direct and indirect comparisons to decide which is longer, heavier or holds more, and explain reasoning in everyday language
(ACMMG006) **AC**

107

Teddies on the Bed

✪ Look at the pictures and tell the story.
One has been done.

There were **6** teddies on
the bed and **2** fell off,
so **4** teddies were left.

_____

_____

_____

_____

_____

_____

_____

✪ Make your own story.

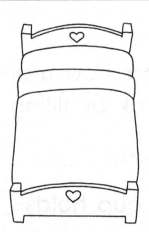

_____

_____

_____

Subtraction (TRB pp. 126–129)
Number and place value Represent practical situations to model addition and sharing
(ACMNA004) **AC**

I'm Hungry

I had this many.	I ate this many.	Now I have this many.

Unit 27 **Subtraction** (TRB pp. 126–129)
Number and place value Represent practical situations to model addition and sharing
(ACMNA004) **AC**

109

Teddy Bears' Picnic

✹ These teddy bears are having a picnic.

Cross out the food to show what they ate.

✹ Write about what the teddy bears ate and what was left over.

_____

_____

_____

_____

_____

Subtraction (TRB pp. 126–129)
Number and place value Represent practical situations to model addition and sharing
(ACMNA004) **AC**

DATE:

STUDENT ASSESSMENT

✦ I had 7 nuts in a bag.

2 nuts fell out. How many do I have left? _____

✦ Complete the table.

I had this many.	I lost this many.	Now I have this many.
(9 pencils)	5	
(7 stars)	3	
(6 bears)	2	
(8 fish books)	4	
(9 apples)	8	

✦ Draw and write you own subtraction or take-away story.

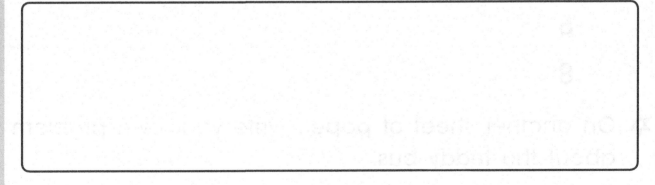

Unit 27 **Subtraction** (TRB pp. 126–129)
Number and place value Represent practical situations to model addition and sharing
(ACMNA004) AC

111

Teddies on the Bus

☆ Use the bus, a dice and some counters to solve the problems.

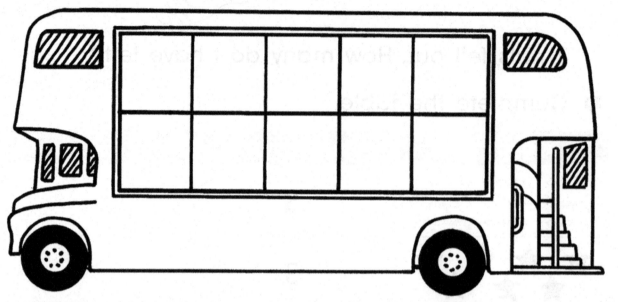

This many teddies got on the bus.	Roll the dice to find out how many teddies got off the bus	How many teddies are left?
9		
7		
8		
10		
7		
6		
8		

☆ On another sheet of paper, write your own problem about the teddy bus.

Jumping Back

✿ A kangaroo is on 8. It jumps back 2 places.
What number is it on?

| | | | | | | | | | | | | |
|0|1|2|3|4|5|6|7|8|9|10|11|12|

8 take away 2 is _____

✿ A kangaroo is on 10. It jumps back 6 places.
What number is it on?

| | | | | | | | | | | | | |
|0|1|2|3|4|5|6|7|8|9|10|11|12|

10 take away 6 is _____

✿ A kangaroo is on 7. It jumps back 3 places.
What number is it on?

| | | | | | | | | | | | | |
|0|1|2|3|4|5|6|7|8|9|10|11|12|

7 take away 3 is _____

✿ A kangaroo is on 9. It jumps back 4 places.
What number is it on?

| | | | | | | | | | | | | |
|0|1|2|3|4|5|6|7|8|9|10|11|12|

9 take away 4 is _____

✿ A kangaroo is on 6. It jumps back 3 places.
What number is it on?

| | | | | | | | | | | | | |
|0|1|2|3|4|5|6|7|8|9|10|11|12|

6 take away 3 is _____

Unit
28
More About Subtraction (TRB pp. 130–133)
Number and place value Represent practical situations to model addition and sharing
(ACMNA004) AC

113

Tenpin Bowling

✵ Use the ten frame and some counters to work out these tens facts.

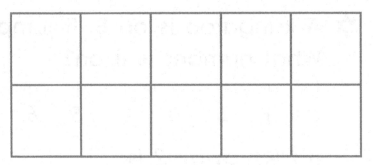

8 counters and _____ spaces make a 10 frame.

3 counters and _____ spaces make a 10 frame.

1 counter and _____ spaces make a 10 frame.

0 counters and _____ spaces make a 10 frame.

✵ Use the tens facts to help solve these problems.

1 🎳 was knocked down.
How many left standing? _____

9 🎳 were knocked down.
How many left standing? _____

3 🎳 were knocked down.
How many left standing? _____

10 🎳 were knocked down.
How many left standing? _____

Unit 28   **More About Subtraction** (TRB pp. 130–133)
Number and place value Represent practical situations to model addition and sharing
(ACMNA004) AC

STUDENT ASSESSMENT

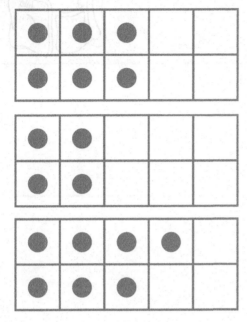

✡ I had 6 and I lost 2.
How many left? _____

I had 4 and I lost 1.
How many left? _____

I had 7 and I lost 3.
How many left? _____

✡ Look at the number line to help you solve these problems.

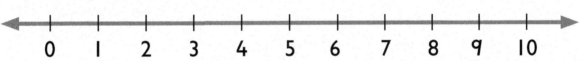

I had 8 and I gave away 3. How many left? ▢

I had 3 and I gave away 1. How many left? ▢

I had 5 and I gave away 2. How many left? ▢

I had 10 and I gave away 6. How many left? ▢

✡ Look at the ten frame. Solve these problems.

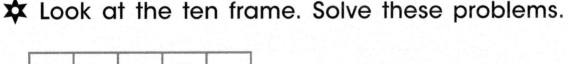

10 take away 8 is _____.

10 take away 3 is _____.

28 **More About Subtraction** (TRB pp. 130–133)
Number and place value Represent practical situations to model addition and sharing
(ACMNA004) **AC**

115

My School Diary

✿ Write about and draw what you do at school during the week.

Monday

Tuesday

Wednesday

Thursday

Friday

✿ My favourite day is _____ because I

_____

More About Time (TRB pp. 134–137)
Using units of measurement Compare and order the duration of events using the everyday language of time (ACMMG007) **AC**

Connect days of the week to familiar events and actions (ACMMG008) **AC**

What's the Time?

✸ Draw what you would be doing at:

 at night.

 in the morning.

 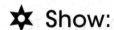 in the afternoon.

✸ Show:

3 o'clock    11 o'clock

5 o'clock   7 o'clock

Unit 29 **More About Time** (TRB pp. 134–137)
Using units of measurement Compare and order the duration of events using the everyday language of time
(ACMMG007) **AC**

Connect days of the week to familiar events and actions
(ACMMG008) **AC**

117

In a Minute

✸ Think about how long 1 minute is.

Draw some things you could do in 1 minute.

✸ Draw some things that would take you **longer** than 1 minute.

 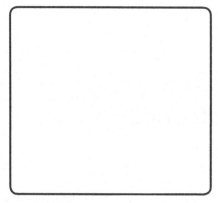 

✸ Do you think it takes **more** or **less** than 1 minute to say "Tiki tikki tembo-no sa rembo-chari bari ruchi-pip peri pembo"? _____

✸ How many times do you think you can turn around while your partner says "Tiki tikki tembo-no sa rembo-chari bari ruchi-pip peri pembo"? _____

Try it with your partner to check.

DATE: _____

STUDENT ASSESSMENT

✸ **What is the first day of the week that you come to school?** _____

✸ **What is your favourite day of the week?**

_____

Why? _____

✸ **Complete the clock.**

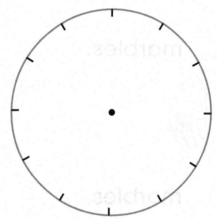

✸ **Write the time for each clock.**

_____   _____

✸ **Show the time on the clocks.**

6 o'clock   **11 o'clock**   **4 o'clock**

Unit
29
More About Time (TRB pp. 134–137)
Using units of measurement Compare and order the duration
of events using the everyday language of time
(ACMMG007) 🅐🅒

Connect days of the week to familiar events and actions
(ACMMG008) 🅐🅒

119

In My Hand

☆ Write how many marbles are in my fist.

_____ marbles.   I have 7 altogether.

_____ marbles.   I have 9 altogether.

_____ marbles.   I have 6 altogether.

_____ marbles.   I have 8 altogether.

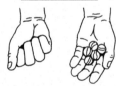

_____ marbles.   I have 7 altogether.

☆ Write how you worked out the answers.

_____

Unit 30 **Addition and Subtraction** (TRB pp. 138–141)
Number and place value Represent practical situations to model addition and sharing
(ACMNA004) **AC**

After the Next Stop

✿ Look at the teddies
on the bus.

How many can you see?

_____

✿ The bus stopped. Now there are 4 teddies
on the bus. Write what happened.

_____

✿ The bus stopped again. Now there are 7 teddies
on the bus. Write what happened.

_____

✿ The bus stopped again. Now there is 1 teddy
on the bus. Write what happened.

_____

✿ The bus stopped again. Now there are 8 teddies
on the bus. Write what happened.

_____

✿ How did you know if teddies got on or off the bus?

_____

Unit **30** **Addition and Subtraction** (TRB pp. 138–141)
Number and place value Represent practical situations to model addition and sharing
(ACMNA004) AC

121

Number Sentences

You will need: two dice

✿ Look at the 2 dice.

Write some number sentences.

4 and 5 is _____   9 take away 4 is _____

5 and 4 is _____   9 take away 5 is _____

✿ Roll the two dice.
Draw what you rolled.

Make some number sentences about the 2 dice.

_____

_____

✿ Roll the 2 dice again.
Draw what you rolled.

Make some number sentences about the two dice.

_____

_____

✿ Draw the 2 dice for this number sentence.

11 take away 5 is 6.

122 Unit 30 **Addition and Subtraction** (TRB pp. 138–141)
Number and place value Represent practical situations to model addition and sharing
(ACMNA004) AC

STUDENT ASSESSMENT

DATE:

✿ How many buttons are in the bag?

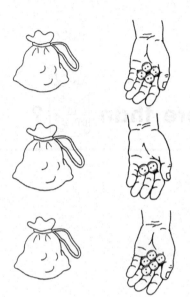

_____ buttons

I have 8 altogether.

_____ buttons

I have 6 altogether.

_____ buttons

I have 7 altogether.

✿ I want to have 4 counters. What do I need to do?

_____

✿ I want to have 7 counters. What do I need to do?

_____

✿ I want to have 5 counters. What do I need to do?

_____

✿ What number sentence can you write using these numbers?

_____

Unit 30 **Addition and Subtraction** (TRB pp. 138–141)
Number and place value Represent practical situations to model addition and sharing
(ACMNA004) AC

123

Maths Challenges

Can you see any numbers around the room?
Copy the numbers onto paper.

Challenge 2

Can you draw a group that has **more than** ?

How many is **less than** 1? _____

Challenge 3

Think of a number that is **more than** 5. Use a paper plate and counters to show the different ways to make it.

Challenge 4

Think of a number that is **more than** 10.
Can you show it in two different ways?

Challenge 5

Make dominoes that
show 7 altogether.

Challenge 6

Draw two groups
that make 6.

Draw two groups
that make 9.

_____ and _____ is 6.

_____ and _____ is 9.

Challenge 7

Draw **two** more than [dice].

Draw **two** more than 1.

Challenge 8

I ate 4 cupcakes. Here are
the cupcakes that are left.

How many cupcakes did I begin with? _____

How did you work this out? _____

_____

Make Your Own Game

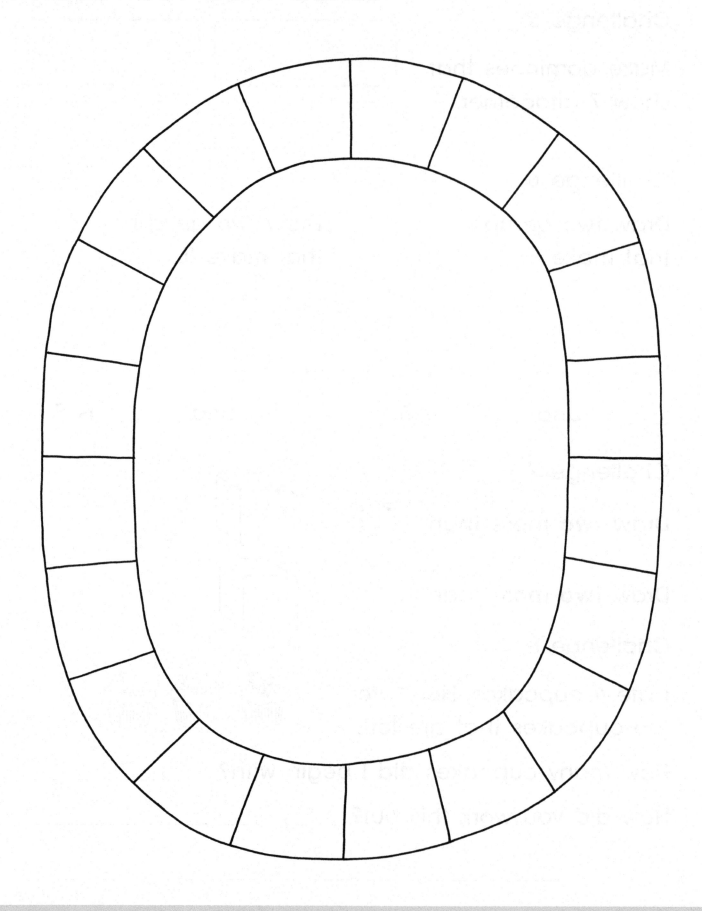

Maths Glossary

Numbers

1	one		11	eleven	
2	two		12	twelve	
3	three		13	thirteen	
4	four		14	fourteen	
5	five		15	fifteen	
6	six		16	sixteen	
7	seven		17	seventeen	
8	eight		18	eighteen	
9	nine		19	nineteen	
10	ten		20	twenty	

2D Shapes

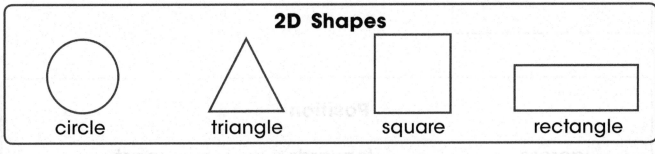

circle   triangle   square   rectangle

3D Shapes

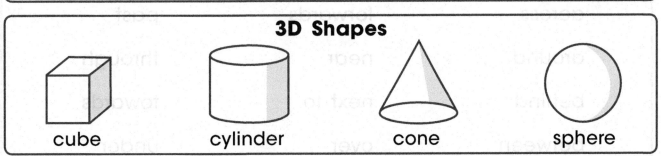

cube   cylinder   cone   sphere

Days of the Week

Sunday   Monday   Tuesday   Wednesday   Thursday   Friday   Saturday

Maths Glossary

Ordinal Numbers

1st	first
2nd	second
3rd	third
4th	fourth
5th	fifth
6th	sixth
7th	seventh
8th	eighth
9th	ninth
10th	tenth

Length

longer

shorter

taller

Mass

heavy

light

Capacity

full

empty

Time

morning

afternoon

evening

day

night

Position

across	forwards	past
around	near	through
behind	next to	towards
between	over	under
in front of		